Educating the Engineer for the 21st Century

# Educating the Engineer for the 21st Century

Proceedings of the 3rd Workshop
on Global Engineering Education

*Edited by*

D. WEICHERT

B. RAUHUT

R. SCHMIDT

*Aachen University of Technology*
*Aachen, Germany*

KLUWER ACADEMIC PUBLISHERS
DORDRECHT / BOSTON / LONDON

A C.I.P. Catalogue record for this book is available from the Library of Congress.

ISBN 1-4020-0096-0

Published by Kluwer Academic Publishers,
P.O. Box 17, 3300 AA Dordrecht, The Netherlands.

Sold and distributed in North, Central and South America
by Kluwer Academic Publishers,
101 Philip Drive, Norwell, MA 02061, U.S.A.

In all other countries, sold and distributed
by Kluwer Academic Publishers,
P.O. Box 322, 3300 AH Dordrecht, The Netherlands.

*Printed on acid-free paper*

All Rights Reserved
© 2001 Kluwer Academic Publishers
No part of the material protected by this copyright notice may be reproduced or
utilized in any form or by any means, electronic or mechanical,
including photocopying, recording or by any information storage and
retrieval system, without written permission from the copyright owner.

Printed in the Netherlands.

# CONTENTS

## PART I. ROLE OF THE GLOBAL ENGINEER IN MEETING THE CHALLENGES OF SOCIETY IN THE 21$^{ST}$ CENTURY

Review of the visions for the Global Engineering Education Workshops     3
*F. W. Stephenson*

The role of the Global Engineer - A European view     7
*T. Hedberg*

The Asian view on the role of the Global Engineer     15
*V. S. Raju*

An American viewpoint on engineering education     23
*P. Kurstedt*

## PART II. INTERNATIONALITY AND INTERDISCIPLINARITY

New demands on engineers - Paths in education leading to professional qualifications     29
*M. Reuber, F. Klocke*

Developing leaders for technology intensive companies - The UNITECH International story     45
*P. Baschera, N. Rickert*

Challenges of a Virtual University Campus: University policy as a consistent system     53
*M. Polke*

Interdisciplinary training of engineers - A challenge between superficiality and overspecialization     65
*Y. J.M. Bréchet*

## PART III. ENGINEERING EDUCATION IN EMERGING ECONOMIES

Present engineering education in India - an emerging economy - and a glimpse of scenario in the 21$^{st}$ century     77
*R. N. Bhargawa*

Lebanon as an engineering educational center in the Middle East     81
*S. M. Chehade*

Science and engineering education in Korea     91
*S. Chang*

## PART IV. EUROPEAN BACHELOR AND MASTER PROGRAMMES

The new engineering bachelor programmes in Italy and at  99
Politecnico di Milano
*R. Negrini*

Restructuring a University level engineering curriculum -  111
A possible response to the Bologna declaration
*A. Kündig*

An American opinion of the European adaptation of the B.S./M.S. degrees  129
*S. S. Melsheimer*

## PART V. DEVELOPING PERSONAL SKILLS TO BE A GLOBAL ENGINEER

The young entrepreneur's experience or: «What can universities do  141
to make of a student a successful Global Engineer ?»
*B. Richerzhagen*

Students' international perspective  151
*M. Püttner*

## PART VI. SUCCESSFUL PRACTICE IN ENGINEERING EDUCATION – PROGRAMMES, CURRICULA AND EVALUATION

Engineering the future of civil engineering in the United States  163
*S. G. Walesh*

Three years of experience with an international graduate program at  173
TU München
*J. Hagenauer, J. Barros, C. Bettstetter, S. Jauck*

Future developments of the European Mineral Programs  181
*H. de Ruiter*

Teaching control engineering to mechanical engineering students  189
by a combination of traditional and modern methods
*H. Rake*

Biomedical engineering education: A case study  191
*M. Mukunda Rao*

Quality management - How to manage education and research  195
in a research institute
*T. Pfeifer, L. Sommerhäuser*

Attracting the next generation of students  201
*D. Brandon*

## PART VII. SUCCESSFUL PRACTICE IN ENGINEERING EDUCATION – EDUCATIONAL CONCEPTS

Learning inventive thinking in the formation of engineer-architects  211
*J. Depuydt*

Acquiring the tools to become a successful engineer in the $21^{st}$ century: 221
Aptitudes and attitudes
*E. Esposito, E. Sigler*

Interdisciplinary curriculum for development of a global engineer  229
*V. Sinha*

A multi-university engineering summer study abroad program  235
*R. L. King, S. Melsheimer, R. Moses*

INTEGRAL - A web-based tool to support learning in  243
interdisciplinary teams
*M. Rötting*

An international collaborative networked venture for mobile  253
communication studies
*V. Sinha, B. H. Walke*

A virtual classroom and more for global education  261
*V. Sinha*

## PART VIII. SUCCESSFUL PRACTICE IN ENGINEERING EDUCATION – UNIVERSITY-INDUSTRY PARTNERSHIP, DESIGN PROJECTS

Northern Arizona University's Design4Practice sequence -  269
Interdisciplinary Training in Engineering Design for the Global Era
*E. Doerry, B. Bero, D. Larson, J. Hatfield*

Further professional education based on working processes -  281
Workflow embedded training in the IT-sector
*M. Rohs, W. Mattauch, J. Caumanns*

Engineering workshops : A multidisciplinary project tightly integrated  289
in engineering education
*J.-C. Léon, P.-M. Boitel, Y. Delannoy*

Capstone design  299
*R. G. Zytner, W. H. Stiver*

Global projects prepare WPI students for the $21^{st}$ century  307
*W. W. Durgin, D. N. Zwiep*

# PREFACE

Upspeeding technological evolution and globalisation characterise today's and future lives of engineers. It is vital for all institutions involved in engineering education to keep pace and to anticipate future needs.

The herein presented collection of papers results from the 3$^{rd}$ Workshop on Global Engineering Education (GEE'3) which took place at Aachen University of Technology, 18 – 20 October 2000. In this meeting more than 150 specialists from 25 countries discussed the topic "Educating the Engineer for the 21$^{st}$ Century".

Which role to attribute to non-technical qualifications? How to integrate ethical aspects in education? Do we have to define international standards in education? What about quality control? What is the potential of new media for knowledge transfer? How to organise lifelong learning for engineers? - These are some of the questions discussed among representatives of industries, educational institutions, politicians and individuals during this meeting.

According to the sessions of the workshop, the book is subdivided into chapters covering the areas "Role of the Global Engineer in Meeting the Challenges of Society in the 21$^{st}$ Century", "Internationality and Interdisciplinarity", "Engineering Education in Emerging Economies", "European Bachelor and Master Programmes", "Developing Personal Skills to be a Global Engineer". Three chapters deal with successful practice in engineering education covering the topics "Programmes, Curricula and Evaluation", "Educational Concepts", and "University-Industry Partnership, Design Projects".

GEE workshops are a joint initiative of Virginia Tech (USA), INP Grenoble (France), EPF Lausanne (Switzerland) and Aachen University of Technology (Germany). GEE'3 was the third workshop in a sequence of similar meetings that took place at Grenoble (1997) and Arlington, Virginia (1998). The next workshop will be held at EPF Lausanne in 2002.

The Editors
July 2001

# ACKNOWLEDGEMENTS

The organisers of this workshop express their gratitude to all sponsors. Their help allowed to invite speakers from developing countries and to create an adequate logistic and social frame for the meeting.

*List of sponsors:*

German Academic Exchange Service (DAAD)

The National Science Foundation, USA

AKAMAI Technologies GmbH, Germany

BMW AG, Germany

Ericsson Eurolab Deutschland GmbH, Germany

RWE Energie, Germany

Stiftung caesar, Germany

ZF Friedrichshafen AG, Germany

Special thanks are due to AKAMAI Technologies GmbH, who made it possible to transmit the Sessions live via the Internet.


## Organization of the 3$^{rd}$ Workshop on Global Engineering Education

*International Program Committee*
C. Arthur - P. Kurstedt - S. Rahman, Virginia Tech (USA)
J.-C. Sabonnadière, INP Grenoble (F)
M. Jufer - B. von Steiger - D. de Werra, EPF Lausanne (CH)
R. Schmidt (Secretary) - W. Weber - D. Weichert (Chair), RWTH Aachen (D)

*Local Organizing Committee*
A. Beyer, Ch. Bischof, C. Bolm, W. Eversheim, K. Heime, K. Indermark,
M. Jansen, R.H. Jansen, F. Klocke, J. Köngeter, R. Kopp, H.R. Maier, M. Nagl,
B. Rauhut, R. Schmidt (Secretary), W. Schröder, H. Wallentowitz,
D. Weichert (Chair), H. Wotruba,

# PART I.

# ROLE OF THE GLOBAL ENGINEER IN MEETING THE CHALLENGES OF SOCIETY IN THE 21$^{ST}$ CENTURY

F. W. STEPHENSON

# REVIEW OF THE VISIONS FOR THE GLOBAL ENGINEERING EDUCATION WORKSHOPS

**Abstract.** The Third Global Engineering Education Workshop begins with a review of the previous two workshops. The partner engineering schools for these workshops are: Virginia Tech, USA; Ecole Polytechnique Fédérale de Lausanne, Switzerland; Institut National Polytechnique de Grenoble, France, and RWTH Aachen, Germany. The partners invited engineering educators and professionals around the world to discuss the changes needed in engineering education to produce successful engineers for the 21st century. This review places the third workshop in perspective relative to the conclusions of the previous two workshops.

The three Global Engineering Education Workshops have defined issues and suggestions in modernizing engineering programs to prepare graduates for the global marketplace. The dates and locations of the three workshops are:

December 10-12, 1997   Institut Supérieure National
                        Polytechnique de Grenoble, France
November 9-10, 1998    Virginia Tech, USA
October 18-20, 2000    RWTH Aachen, Germany

A fourth partner, Ecole Polytechnique Fédérale de Lausanne, Switzerland will host the fourth workshop in 2002.

The first global engineering education workshop set the focus for the following years. The 1997 workshop was a pilot to discuss the issues of preparing engineering graduates for the global workforce. The one-track agenda and invited attendees encouraged a meaningful discussion among all the participants. Academic, industrial, and government speakers participated in the 2.5 day event. The first global workshop identified two goals:

Goal 1 -   New directions for action involving new roles and rewards for faculty, new dimensions for graduate education incorporating technology, new common focus on learning or outcomes rather than inputs such as teaching, and sharing new approaches to educating students.
Goal 2 -   Reinforce international cooperation among participants as a basis for understanding through experiencing the culture of the four partners' academic campuses, and informal conversations flowing from formal presentations throughout the workshop.

The first workshop identified issues forcing change. The summary of the first workshop in Grenoble in 1997 was presented by Charles Steger, President of Virginia Tech. While predicting the future is difficult, changes in the global economy require change in communication technology and the nature of work. Many jobs may become part time and/or involve many career changes. Another issue forcing change is the quantity and velocity of global capital flow. The new and highly efficient market mechanisms have changed the structure of the world economic system.

The carrying capacity of the planet is uncertain. We face many questions. How do we develop sustainable design strategies? How do we relate a global perspective to local action? How do we balance economic growth with environmental protection? How do we achieve equity between developed and developing countries?

Because of these issues driving change, higher education must discuss what body of knowledge is appropriate for global engineers. How do we integrate knowledge from the present with the past? How do we certify/validate global curricula? How will new learning strategies be developed and integrated into our universities? How do we facilitate cross-cultural experiences?

Finally, the first workshop in 1997 asked: Will networks and increased university cooperation produce needed synergy? Will students gain international experience through internships in real or virtual corporations? Who will gain from the value added by new, collaborative international initiatives?

The Second Global Workshop was hosted in Arlington, Virginia, USA, just outside Washington D.C. in November, 1998. Eleven countries were represented: England, France, Germany, Japan, Malaysia, Northern Ireland, People's Republic of China, Romania, Singapore, Switzerland, USA. Thirty universities were represented, 30 from the U.S. and 12 from the participating countries. Corporate sectors were represented by manufacturing, information technology, and space technology.

The second workshop focused on destinations and directions. The topics discussed included:
    Requiring the global engineer
    Developing global faculty
    Demonstrating education tools for global engineering skills
    Molding the global graduate and post graduate
    Providing resources for the global education environment

As a group of engineering educators, researchers, leaders, practitioners, and government officials, the participants were concerned about future directions in global education. If knowledge generation is so critical, why are universities in the U.S. and Europe suffering from reduced budgets? Are demographic shifts in advanced industrial societies driving resource allocation priorities?

For the third workshop, hosted by RWTH Aachen in October, 2000, the discussions become more specific. Educating the Engineer for the $21^{st}$ Century includes discussions on the issues of International, Interdisciplinary, and Information Technology. Specifically, the discussions are Successful Practice in Engineering Education, Developing Personal Skills to be a Successful Global Engineer, Educating the Global Engineer in Emerging Economies, and the Concept and Accreditation Capability of New Bachelor and Master Programs in Europe. The third workshop can address unanswered questions and issues from previous workshops. Workshop One: Why do faculty participate in global education activities and how do we reward such participation? Workshop Two: How do we provide resources for the global environment and how do we generate resources for global education? Workshop Three: How do we make higher education (in general) and global education (in particular) a more important issue for the general public?

This third workshop includes scientific visits to research institutes and enterprises before and after the workshop to promote collaboration between U.S. and European researchers. The workshop is video streamed on the internet and we have people around the world listening and watching us. We look forward to another stimulating workshop.

*F. William Stephenson*
*Virginia Tech*
*College of Engineering*
*Blacksburg, Virginia, USA*

T. HEDBERG

# THE ROLE OF THE GLOBAL ENGINEER

*A European View*

**Abstract.** The main problem in Engineering Education today is not to define the qualifications needed for the future global engineer. There is a very degree of consensus concerning the aims and objectives of the education. The problem is how to attain these ambitious goals. In the paper the author argues in favour of continuous professional pedagogical development and underlines the importance of the "hidden curriculum" - the attitudes and values transferred to the student during the studies.

Qualified "global" engineers will no doubt be needed in the future, even more than today. We are rapidly moving into a global knowledge-based society, high-tech products become increasingly important and every country is concerned about the international competitivity of its industry. All engineers will in the future work in an international context, cooperate with people from other countries and many will spend part of their career in other countries.

Global Engineers are needed to advance technology in harmony with the evolution of society and with the dreams and wishes of its citizens. It's part of their responsibility to apply technical skill and competence to the many urgent problems our societies are facing - environment, energy, lack of food, poverty, lack of water, disease… .

But what does all this exactly mean? What are the consequences for our universities and our Engineering Education? What kind of engineers will be needed in the future? How should the engineering education of today be modified to make us better prepared for the future?

The engineers we educate today will be active 40 years from now. To predict in any reasonable detail the development over such a long period is of course an impossible task. However, I don't think we have to know exactly what will happen. If we manage to educate an engineer as we think they should be educated now then I believe that we have found a good solution to our problem. A considerable effort has been made in many countries to determine what an ideal engineer should master and a variety of solid studies and reports full of good suggestions have been produced. There is fact a very high degree of consensus concerning the new demands on the engineer.

The engineer of tomorrow should of course have a high technical and scientific competence. In addition he or she should be able to communicate in his or her native language, in English and preferably in at least one more foreign language. Cross-cultural communication skills are needed as well as managerial skills and ability to work in teams. He or she should have a deep understanding of ethical and environmental issues, be broadminded, innovative, imaginative and creative, well versed in humanities and have a deep understanding of the relationship between technology and social development. He or she should be curious, have a good common sense, be willing to learn and able to take responsibility. This high ideal is something that we might call a "Renaissance Engineer".

As an example the French Commission des Titres d'Ingénieur insists that none of the following elements may be missing in a curriculum to be accredited by the Commission:

- a thorough education in the basic sciences,

- a complete education in the general techniques of the engineer, including the mastery of complex systems,

- a sufficient training in the main fields of the chosen specialisation,

- a general education comprising foreign languages, economic, social and human sciences, communication and an introduction into the ethical reflection about the engineer's role,

- a training for life and for the problems of the enterprise, also in their international dimension.

The fundamentals of quality, hygiene, security, environment and intellectual property must be part of the curriculum. Documents from the British Engineering Council give the same picture: An engineer should not only be technically competent but also "market conscious, commercially adept, environmentally sensitive and responsive to human needs".

The official Swedish text says:

- ...

- obtained knowledge and skills needed for the development of products, processes and work environment taking into account human conditions and needs and the aims of the society concerning social relations, sustainable development and economy.

- ...

A recent Swedish study is a bit more elaborate. It argues that an engineer should have technical competence, social competence, a holistic view, be aiming at results, have administrative skills, be quality minded, creative and flexible, be anchored in reality, be hungry for knowledge, have a managing ability, have good knowledge of foreign languages, possess an ethical competence and be responsible.

Most European educators and industrialists would agree on the list of competencies defined by ABET:

"Engineering programs must demonstrate that their graduates have:

- an ability to apply knowledge of mathematics, science, and engineering,

- an ability to design and conduct experiments, as well as to analyse and interpret data,

- an ability to design a system, component, or process to meet desired needs,

- an ability to function on multidisciplinary teams,

- an ability to identify, formulate, and solve engineering problems,

- an understanding of professional and ethical responsibility,

- an ability to communicate effectively,

- the broad education necessary to understand the impact of engineering solutions in a global and societal context,

- a recognition of the need for, and an ability to engage in life-long learning,

- a knowledge of contemporary issues,

- an ability to use the techniques, skills, and modern engineering tools necessary for engineering practice."

These statements and corresponding ones from other countries are similar in spirit although the wording and emphasis might be slightly different. I am personally strongly convinced that the goals formulated above and in other similar documents are worthwhile. Engineers with these qualifications will be able to meet challenges not only of today, but also of tomorrow. Engineers with these competencies would be well prepared to assume the Rôle of the Global Engineer.

However, what we actually see at our universities does not always correspond to all these ambitious goals; certainly not in my country and probably not always in others. The central question is therefore not *what* the aims and objectives are, or how they should change. The question is if and *how* we can attain them. There is obviously no single simple answer to this question. We must perhaps even accept that not all our engineers will attain these goals. Some engineers will be local, want to be local, do a good job as "local" engineers, satisfy a local need and be happy with that. In their career they may primarily seek positions of a technical character in the narrow sense. But this is certainly no reason why we should loose our aims out of sight.

The graduate engineer is a product of his university. If we want open-minded, innovative and creative engineers willing to experiment and take risks, then we must have an open-minded, innovative and creative university able to experiment and take risks. If we want our students to be able to function on multidisciplinary teams, to understand their professional and ethical responsibility, to be able to communicate effectively in many languages, to understand the impact of engineering solutions in a global and societal context, to recognise the need for and engage in lifelong learning, to know about contemporary issues, then we must also have a faculty with the same abilities.

Teachers transfer attitudes and values to the students and the attitudes and values that are transferred during the important university years are of utmost importance. It's obvious that the professional character of engineering education to a large extent depends upon this. Through contact with his teachers the student gets familiar with - and starts to adopt - the language, attitudes, values and usage of the field and profession he has chosen. When we want to change how our future engineers are educated then we must not ignore this part of the curriculum, the "hidden curriculum". It must be discussed and analysed in an open way so that teachers and students become aware of these attitudes and values. And those responsible for the education must be reasonably convinced that the hidden curriculum indeed does contribute to the creation of those "renaissance engineers" we want. If not, then something has to be done.

All this calls for a systematic and continuing professional development of the teaching staff. It is a strange and striking fact that universities, although deeply involved in education, believing in the value if education and advocating lifelong learning for others, much too often seem to neglect the lifelong learning of its own employees, of its own faculty. Of course there are many obstacles - tradition, scarce resources and lack of time - but these obstacles are common to all organisations and in no way specific for universities. They have to be overcome in universities as well as in industry.

Research, which ideally should be performed by every teacher, is one form of on the job training and a part of a lifelong learning. However, this is far from enough. First, not all teachers are actively involved in research. Secondly, the research field is too often quite narrow, compared to the broad and rapidly evolving field of engineering science for which the teacher may be responsible.

Some universities have made admirable efforts to improve the quality of teaching, but there is still much to be done when it comes to the pedagogical training of university teachers. The role of the teachers is changing as we change emphasis from teaching to learning and move towards more student centred methods. The teacher should also be aware of new pedagogical concepts, new theories of learning and new methods of teaching. The teacher has also to reflect upon his profession. Universities therefore must promote the pedagogical development of its teaching staff, experienced as well as less experienced teachers.

There are many interesting pedagogical experiments going on all over the world; some successful and some probably less successful. Universities are however often too slow to study and take advantage of experience obtained elsewhere and should be much more open to innovation in this domain.

Much has also been said about university-industry cooperation. The focus is normally on the benefit for industry; how knowledge should be transferred from universities to industry, how the creation of new research based firms be should encouraged, how universities should be involved in life long learning of professional engineers etc. Co-operation could also mean research projects conducted together with an industrial partner and sometimes entirely or in part financed by that partner.

Much less has been said about the benefits of industrial cooperation for the engineering education at universities. Engineering teachers are not always aware of the rapid changes in industry. An active participation in industrial cooperation will facilitate for them to keep abreast of these changes and is therefore another element of the lifelong learning for a professional teacher. This is particularly true for those who lack a personal first-hand industrial experience. Universities should also encourage and facilitate its faculty members to spend time in industry as a natural part of their career plan.

Hiring and promotion are still in most countries essentially based upon performance in scientific research. If we want to educate renaissance engineers then we should pay more attention to other merits, pedagogical, industrial and qualifications such as those we want to see among our students.

All universities are continually reshaping their engineering curricula, but it is certainly a most difficult task to combine all the necessary engineering and scientific fundamentals with new demands on social, economic, managerial and communication skills within a restricted schedule and time. This task is certainly not facilitated by shrinking recourses and increasing student numbers. It is however something that has to be done. An introduction of more active and student centred learning methods is needed, as case studies and project-organised teaching. Problem-based learning also offers a possibility and here there are some successful examples (e.g. Aalborg, Denmark) that can serve as models. But there can exist no single model for the engineering curriculum; on the contrary, diversity is important.

Learning is of course more important than teaching, students must learn how to learn and IT or ICT and "flexible learning" will have an impact on our engineering education. I refuse however to believe that IT will solve all our problems. There are in fact some dangers that we must be aware of. Internet is certainly useful, but we must not confuse information with knowledge. Hands-on experimental work is important. Computer simulations are valuable, but if we replace all laboratory work by some kind of virtual reality we lose something important. It is even becoming increasingly important to have experimental work done at the universities, as the technology encountered in our daily life is becoming less and less transparent.

It's not enough to have first class curricula and first class faculty. We also need good students and many good students. A major worry in many European countries is the fact that technology and science do not today attract enough young men and women. In Sweden today only about 20% of students leaving the compulsory lower secondary school (at the age of 16) choose science or technology options in upper secondary school, i.e. high school (ages 16-19). Not only engineering schools recruit from this group. Some become teachers, medical doctors, study science at universities etc. and some choose non-scientific careers. The number of young people following the scientific high school programme is therefore not even high enough to satisfy the need of today and risks to make the attempts to expand university education in science and technology futile.

This problem is not unique to Sweden; in fact our country might be in a better position than many other industrialised countries. It is however beyond the scope of this paper to analyse the underlying factors behind this disenchantment with technology and science, why there is a falling enrolment to science at high school level in many European countries.

Many universities are making admirable efforts to promote the interest for technological and scientific studies. But the problem goes far beyond what individual universities can solve. It calls for large-scale national efforts involving government, universities, industry and schools. The Swedish government has in fact initiated an ambitious national programme, called the NOT-project, to address and redress the situation. The first phase of the project has been running from 1993-98 and is followed by a similar five-year programme. The aim is to reach many and different target groups. The strategy has been to influence teachers, headmasters, parents and other adults who are important to children of varying ages. An array of school improvement tools is used - new teaching materials, newsletters, conferences, in-service training of teacher trainers, and cooperation between schools and Science centres. A popular book on science has been produced and spread widely. A large mass media campaign has been carried out.

A special attention is given to the recruitment of female engineering students. Their share has increased gradually over the years and today it makes up about 28%. The goal is to increase that even further. More female students are needed if we want to increase the total number of graduates in engineering. Male and female engineers also have different experiences from life. Female engineers can therefore enrich technology by bringing in another perspective.

The falling standards of entering engineering students are also a major concern in many European countries, including Sweden. This is particularly obvious in mathematics, but complaints are also heard from other areas. Part of the decline is certainly due to the fact that recruitment has increased over the last years. This increase cannot explain, however, all the observed changes. Changes in the high school curriculum and insufficient teacher's training might be part of the explanation. More important factors are probably changes in attitude, behaviour and values of the younger generation. Universities will most certainly have to adapt to the new situation and be prepared to receive students with more varying skills and knowledge than they were used to in the past.

Finally a few words about what might be call the "Ingenieurmässigkeit" - the specific professionalism of the engineer. Part of it is created by the hidden curriculum and students get an introduction to this through contact with engineers. It's therefore very important that at least some faculty have a qualified personal industrial experience. Unfortunately this is not always the case and it at least in our country it also seems to become less common. To some extent the problem can be overcome by the use of part time adjunct staff and through industrial cooperation. The importance of design must also be stressed. Design is the core of all engineering activity and some experience of design of a product, a system or a process should be included. An industrial placement period, preferably both a blue-collar experiences on the shop floor, and a period as an apprentice engineer should also be an integral part of the engineering education.

As you know we have in most European countries essentially two basic types of engineering education; a long-cycle - often 5 years - and more vocationally oriented short cycle, often 3 years. The problem of academic drift is clearly seen in many cases, a tendency to make in particular the short cycle education more academic and less professional, less "ingenieurmässig". I deeply regret this tendency and I do not think that what happens is in the interest of neither students nor society.

I have said nothing about the Bologna declaration and proposed the 3+2+3 system. It might work in Europe and after all it seems to work elsewhere. I have no particular concern about the 5-year Dipl.-Ing./Magister/Master/Ingénieur Diplomé/civilingenjör level. On the other hand I'm afraid that the introduction of such a "Bologna system" in engineering might amplify the academic drift of the short cycle education.

I hope that the university of tomorrow will be a student-centred university, a university with student-centred learning methods and a university listening to its students. Students want to discuss and participate, individually or through their representatives, in the decision process concerning how teaching and learning should be done, and they should be encouraged to do so. Continuous quality improvement of a university requires feedback from the students through formalised and informal channels and the active participation of students in the university gives training for democracy. The degree of student participation is thus a measure of the quality of a university and a prerequisite for quality improvement.

The young generation is changing and students today want to define their individual life projects and their own learning objectives. Students are adults and should be treated as such. They carry the main responsibility for their education and professional development. The university is just one short part of their lifelong learning, and, in this respect, there is no difference between this part and the part carried out after graduation.

*Torbjörn Hedberg*
*Luleå University of Technology, Department of Mathematics, Luleå, Sweden*
*President, the European Society for Engineering Education (SEFI), Brussels*

V.S. RAJU

# THE ASIAN VIEW ON THE ROLE OF THE GLOBAL ENGINEER

**Abstract.** While I have been asked to present the Asian view, understandably the views expressed here are more based on the Indian Experience. In every respect, probably Asia is much more diversified than any other continent. A division in to developed and developing countries may be another way of looking at different perspectives. From this angle, Asia still belongs more to the developing world, baring Japan, Korea and Singapore, inspite of some spectacular economic growth rates in the recent past. Inspite of the variations, the points made in this presentation, I believe will apply to a majority of the population in Asia.

## 1. THE 21$^{ST}$ CENTURY CHALLENGES

Some of them are:
- (a) Population explosion
- (b) Depleting natural resources - example: energy, water, materials
- (c) Environmental degradation - air, water, soil, noise pollution, solid and liquid wastes
- (d) Widening of inequalities in wealth and access to resources (between countries and within a country)
- (e) Societal tensions, endemic youth unemployment
- (f) High impact of technology (example: information technology, biotechnology, genetic engineering) on society, commerce, education, entertainment, life styles
- (g) Unipolar world
- (h) Demand for mass education for improvement in living standards (goods and services)
- (i) Globalisation

While all the above are issues for the entire world, these are more acute for the developing world, to which most part of Asia belongs.

In addition, the crucial issue for Asia is extremely rapid urbanisation. For example in 1975 the world population totalled 4 billion people with 34% of these living in an urban environment (0.81 billion in cities of the developing world and 0.73 billion in the cities of the developed world). By 2025 the projections are a world population of 8.29 billion people with 61% of these living in urban environments. This translates to 4.03 billion in the massive urban sprawls of the developing world and the hardly increased 1.04 billion living in the cities of the developed world (P. Annez, „Livable Cities for the 21$^{st}$ Century", Siemens Review, vol.62 nos.3-4, pp. 5-11, June/July 1996).

Such a dramatic growth in city population in the developing world imposes tremendous pressures on all of the areas of concern listed above. Survival followed by wealth and employment in an urban environment implies total dependence on infrastructure development for transport of essential goods, water and food, public transport, sanitation, etc. In addition, to meet the demands of international trade and over all development, massive efforts are needed to develop highways, railways, seaports, airports, waste management systems, etc.

## 2. DESIRABLE CHARACTERISTICS OF XXI CENTURY ENGINEERS

This has been a subject of discussion in several conferences and workshops. A typical summary is given below, which may be considered as a global view:

*Table 1. The desirable characteristics of XXI century engineers*

| TRADITIONAL ATTRIBUTES | XXI CENTURY ADDITIONAL ATTRIBUTES |
|---|---|
| ❖ Problem-solving abilities | ❖ Learnability : learning to learn, on one's own |
| ❖ Analytical skills | ❖ Strong desire for life-long learning - continuous education |
| ❖ Communications skills - oral, written, graphic | ❖ Ability to work in a team |
| ❖ Ability to relate to practical aspects of engineering | ❖ Exposure to commercial disciplines |
| ❖ Inter-personal skills | ❖ Creativity and innovation |
| ❖ Management skills | ❖ Integrative skills |
| ❖ Decision-making skills | ❖ International outlook |
|  | ❖ Ability to deploy IT |
|  | ❖ Ability to work at interfaces between traditional disciplines |
|  | ❖ Commitment to sustainable development |

## 3. EXCEPTIONALLY HIGH DEMAND FOR QUALITY HIGHER EDUCATION IN ASIA

The percentage of youngsters going into higher education system in developing Asia is relatively low. At the same time the rate of growth is phenomenal.

*Table 2. Indian example: Enrolment in higher education, 1901 - 1997, per hundred thousand population*

| Year | 1901 | 1917 | 1951 | 1961 | 1971 | 1981 | 1991 | 1997 |
|---|---|---|---|---|---|---|---|---|
| Enrolment per 100,000 | 12 | 23 | 48 | 126 | 360 | 402 | 546 | 613 |

*Source: Higher Education in India: Vision and Action, UNESCO World Conference on Higher Education in the Twenty-First Century, Paris 5-9, October 1998.*

The demand for technical education is growing at a much faster rate than for the other disciplines. For example, the situation in India: Every year nearly one million youngsters are aspiring to enter an engineering degree program. They write several entrance examinations, to get into one of the engineering institutions both at national level as well as regional level. In addition to preparing and writing the school examinations, they will have to go for special coaching for these entrance examinations. For the 6 Indian Institutions of technology, 1,300,000 aspirants write the joint entrance examination just for 2500 places. Because of the intense competition, the aspirants specially prepare for two or three years for these entrance examinations. In the process, they become quite competent. Internationally benchmarked, I estimate that 80,000 of them will be of a calibre to be able to go to best institutions anywhere in the world and successfully complete the program.

While similar data is not readily available for the other Asian countries, my estimation is that the situation is no different, though the degree may vary.

## 4. EXTREME ISSUES OF QUALITY OF THE EDUCATIONAL PROCESS

Because of paucity of resources, even to ensure the quality of institutions in existence for some time is very difficult. In addition, the rate of expansion is so large that simply the resources are not available for ensuring quality. Along with financial resources, availability of the human resource, namely quality faculty is very acute. Even with well endowed institutions, faculty positions are not attractive because of the large salary differentials that exists with the industry. Equally crucial is the fact that not many are opting for doctoral programs in these countries. The Indian experience is that the very best graduates at Bachelors level either go to an industry, management programs or in some cases even to civil service. Those few who want to pursue graduate programs in their respective disciplines would prefer to go to one of the developed countries, in particular United States.

In earlier years some of these people who did their doctorates abroad used to come back to the home country and take up assignments as faculty and researchers. Unfortunately, that number is decreasing because of the international demand for quality technical manpower. Yet another factor is the establishment of R&D centres by multinationals in Asian countries. While this is most welcome in an over all sense, it is immediately impacting the educational institutions as some of the good faculty are migrating to these R&D centres and some who otherwise join the academic institutions are opting for these centres.

## 5. GLOBAL DEMAND FOR ASIAN MANPOWER

It is an accepted fact or a reality that due to globalisation there is over all mobility and technical manpower from Asia are in demand all over the world, in particular North America and to some extent Australia. There are two aspects to this:
1. The Asian countries have to train sufficient number of engineers to meet their own demand as well as global demand.
2. The education process has to be in match with the global systems to provide mobility to the graduates.

In a broader sense also the educational processes all over the world, both in the developed and developing countries are tending to be similar or comparable. The variations are essentially due to lack of resources.

## 6. THE EXPLOSION IN INFORMATION TECHNOLOGY (IT) OPPORTUNITIES AND ITS IMPACT ON AVAILABILITY OF MANPOWER IN TRADITIONAL DISCIPLINES

There is a very large unfulfilled demand for IT professionals both globally and in the Asian countries. The University system is not in a position to train sufficient number of IT graduates. Because of the capacity of the IT industry to pay much higher salaries, persons qualified in the other engineering disciplines like Electrical, Mechanical, Chemical, Civil are migrating to IT. The result is bright and sufficiently trained engineers are not available to take care of the needs of infrastructure projects like roads, ports, railways, manufacturing facilities, etc.

## 7. POSSIBLE SOLUTIONS

Main Objective:
1. To create and provide high quality, globally benchmarked, technical educational opportunities for aspirants from Asia.
2. To meet the manpower needs of the respective country, keeping in view the regional and global demands.

### 7.1. Networking

For optimum utilisation of scarce resources and for mutual benefit

(a) among academic institutions in same country,
(b) among the institutions in the region,
(c) among institutions across the globe.

## 7.2. Bilateral and Multilateral Cooperation

### 7.2.1. Institution building

Example: An outstanding example of this cooperation is the establishment and development of the Indian Institutes of Technology (IITs) in India, 5 of them, rated among top 10 in Asia Pacific today. Theses were joint ventures between India and different countries namely Germany, U.K., U.S.A. and Russia. The partnership with Germany was for IIT Madras and our host, the Technical University Aachen and several other Technical Universities in Germany have played very crucial role in building up various facilities by sending their faculty and training faculty from IIT Madras in Germany.

While the benefits for the host country are obvious, there are many advantages for the partner country as well. This include flow of quality graduate students, U.S. Universities have immensely benefited in this aspect. Several thousands of graduates from IITs in India have gone over to U.S. to do graduate programs and subsequently participated in a very substantial way in the technological and industrial development of that country.

### 7.2.2. Faculty Development - Top most Priority

This would benefit the partner institutions as well by way of quality graduate students.

### 7.2.3. Faculty Exchange

This will give an exposure to the faculty of the partner institution to a different environment, which can also be very enriching.

### 7.2.4. Students Exchange

Benefits are well recognised. For industrial cooperation or collaboration between Asia and the developed world of Europe and North America, it is very essential to have their engineers exposed to each other countries.

### 7.2.5. Joint Courses / Degrees

Various forms are available for this.

### 7.2.6. Distant Education and Multimedia

This is going to play a very crucial and vital role in meeting the challenges of education. In essence, in some disciplines like IT, this mode is possibly the only hope to meet the challenges.

### 7.3. Links between Academia and Industry

While its importance is well recognised, baring few institutions this is at very low level or non-existent in many countries of developing Asia. The IITs in India have made some spectacular progress during the last 5 to 10 years.

### 7.4. Government and Private Partnerships

Till very recently most of the higher educational institutions in Asia have been fully funded by Government. However it is now realised that Government does not have sufficient resources to fund the rapid expansion that has become necessary to meet the demands. There are number of private institutions, which have come up recently. However quality is a major concern. Ideally a partnership between the two would yield better results and also speed up the process of capacity building.

### 7.5. Alumni Participation and Support

For a very long time, North American Universities have been keeping close touch with their Alumni and have in the process immensely benefited. Alumni have made substantial contributions in terms of finances and also in many other ways by participating in management, in providing links between their organisations and the Alma Mater. The Asian Universities should do the same.

In India, during the last 4 to 5 years, the Indian Institutes of Technology have made determined efforts in this direction. The Alumni, in particular those who are in United States as well as Entrepreneurs in India have taken great initiatives. Already for the 5 IITs contributions to the tune of about 200 million U.S. dollars have been received. Alumni have committed to raise between U.S. dollars 500 million and a billion in the next 10 years. Similarly the Indians in the United States, in particular the Alumni of the IIT system, have plans to raise one billion dollars to establish 4 or 5 Global Institutes of Science & Technology in India. They have announced that 300 million dollars have already been raised for this purpose.

## 8. SUMMARY

1. There are many 21$^{st}$ Century challenges. While these are issues for the entire world, they are more acute for the developing world to which most part of Asia belongs. In addition, the crucial issue for Asia is extremely rapid urbanisation. For survival, massive build up of physical infrastructure is a prerequisite.

2. With the focus on greater need for infrastructure build up in Asia, the desirable characteristics of the 21$^{st}$ century engineers in Asia are similar to the rest.

3. The demand for quality technical education in Asia is exceptionally high.

4. There are extreme issues of the quality of educational process, because of paucity of resources both human and financial.

5. The demand for Asian manpower is global, indicating the need for training manpower to meet own demands as well as global demands.

6. The explosion in Information Technology (IT) opportunities are adversely impacting on the availability of quality manpower in traditional disciplines of engineering.

7. Possible solutions to create globally benchmarked high quality technical educational opportunities in Asia:

   a. Networking among academic institutions in the country, in the region and across the globe.

   b. Bilateral and multilateral cooperation for institutional building, faculty development, faculty exchange and students exchange, joint courses and degrees, use of distance education and multimedia techniques, links between academia and industry, Government and Private partnerships and finally Alumni participation and support.

*V.S. Raju*
*Indian Institute of Technology Madras*
*Ocean Engineering Centre*
*Madras, India*

P. KURSTEDT

# AN AMERICAN VIEWPOINT ON ENGINEERING EDUCATION

**Abstract.** The global economy is greatly affected by the number of start-up organizations in both new and established companies. U. S. engineering schools are addressing the new entrepreneurial skills their graduates must know to be successful in the global economy. Examples are given of engineering schools, private foundations, and companies working together to present information and experiences to students to develop a larger pool of entrepreneurs. Some of the information and experiences are integrated into the curricula and some are offered as extracurricular activities and internships. There are many questions to be answered about teaching this information to students—who, when, where, and how.

In the United States, the new economy is driven by information technology. The internet provides immediate access to global markets and partners. These new markets and the use of technology to access those markets have created thousands of start-up companies in the United States. The skills needed by employees in start ups have not been taught in engineering schools. In several schools, the faculty are attempting to teach entrepreneurial skills within the engineering curriculum. In the past, most faculty have not developed entrepreneurial skills themselves. The question becomes, who will teach engineering students and faculty to be entrepreneurs in a global market?

In the 21$^{st}$ Century, employers want to hire engineers with global market and finance savvy. The accrediting agency for U.S. engineering schools, ABET, recognized the need for engineering graduates to demonstrate more than technical skills. The *ABET Engineering Criteria 2000* includes the following in *Criterion 3. Program Outcomes and Assessment.*

Engineering programs must demonstrate that their graduates have
- a) an ability to apply knowledge of mathematics, science, and engineering
- b) an ability to design and conduct experiments, as well as to analyze and interpret data
- c) an ability to design a system, component, or process to meet desired needs
- d) an ability to function on multi-disciplinary teams
- e) an ability to identify, formulate , and solve engineering problems
- f) an understanding of professional and ethical responsibility
- g) an ability to communicate effectively
- h) the broad education necessary to understand the impact of engineering solutions in a global and societal context
- i) a recognition of the need for life-long learning
- j) a knowledge of contemporary issues

k) an ability to use the techniques, skills, and modern engineering tools necessary for engineering practice.

The requirement for engineering graduates to function on multi-disciplinary teams, communicate effectively, understand the impact of engineering solutions, a knowledge of contemporary issues, and an ability to use the techniques, skills, and modern engineering tools are all needs of entrepreneurial organizations.

According to Forrester Research, by 2004, 7% of U.S. retail sales will be on-line purchases. The global internet economy could reach $6.9 trillion. By 2003, global business-to-business sales could reach $1.8 trillion. Everyday, there are new predictions in the media about how technology will impact national economies and our cultures. Higher Education is attempting to meet the needs of this new economy requiring entrepreneurial skills.

According to the E-Commerce Learning Center at North Carolina State University, http://dmoz.org/Business/E-Commerce/Education, presently there are 39 degree programs and 30 research centers throughout the world, although focused in the U.S., related to E-Commerce. This trend in responding to the business world seems to be growing. Stanford University hosted a discussion in October, 2000, entitled, *Roundtable in Entrepreneurship Education for Engineers*, http://www.stanford.edu/group/stvp/ . This discussion highlighted the interest and struggles engineering schools are encountering as they try to modernize the engineering curricula to meet the needs of the New Economy.

Offering new courses or modernizing degree programs seems appropriate. However, a key question will be who is qualified to teach these entrepreneurial subjects? In many cases, engineering programs are partnering with business schools to offer these courses (Cornell, Carnegie Mellon, MIT). At other engineering schools, they offer entrepreneurial topics within the school of engineering (Georgia Tech, Stanford, U of Colorado-Boulder, Virginia Tech). Faculty who have generated their own start-up companies often teach these courses. Successful entrepreneurs who are alumni of the engineering schools often serve as adjuncts teaching these topics.

Industry is offering extracurricular entrepreneurship programs for student entrepreneurs. Arthur Anderson is a sponsor for StartEmUp, http://www.startemup.com . StartEmUp creates alliances between students and venture capitalists. In 2000-01, StartEmUp hosted JumpStart2K Campus Conferences for university competitions in business plans of student-directed start-up ventures in 13 cities for 30 universities. The winning student business plan received $100,000 and continued assistance to continue efforts toward a successful commercial enterprise.

There are also successful private foundations such as the Kauffman Center for Entrepreneurial Leadership, http://www.emkf.org , funding education programs aimed at entrepreneurial engineering students. The Kauffman Center believes the world economy will rely more on entrepreneurial philosophies and practices. The provide materials, scholarships, internships, and fund projects that promote entrepreneurship among students. One example of the Kauffman Center materials is the article by Marilyn L. Kourilsky, Vice Present for the Kauffman Center for Entrepreneurial Leadership entitled *Entrepreneurship Education: Opportunity in*

*Search of Curriculum*, Business Education Forum, October, 1995. One of the programs Kauffman Center funds is the Entreprep for talented high school students. This program provides internships in start ups and provides college scholarships. The Kauffman Entrepreneur Intern Program provides internships in start-ups for college students.

Engineering colleges need industry input to design and teach entrepreneurship programs. The internet can allow colleges to use experts from around the world to participate in teaching. Practicing entrepreneurs can teach the skills for success that traditional engineering teaching and research faculty of this generation may not be able to offer. Industry will want to hire graduates with experience in start ups, so internships in an entrepreneurial setting will be very important as part of the professional preparation of engineers. These internships can also provide an introduction to creativity in products, processes, markets, and financing. This creativity will drive success in the New Economy.

The response of U.S. engineering schools to teaching entrepreneurial skills is essential for our companies and organizations to remain globally competitive. The questions which need to be answered are related to vision, courage, will, knowledge, and responsibility. These questions will require the collaboration and cooperation of both higher education and industry to produce the graduates for the 21$^{st}$ century.

## REFERENCES

1. McGraw, Dan, Getting Down to E-Business, Preparing Students for Entrepreneurial Roles in an Internet Economy, *ASEE prism*, October, 2000.
2. Websites as listed in text.

*Pamela Kurstedt*
*Virginia Tech*
*College of Engineering*
*Blacksburg, Virginia, USA*

# PART II.

# INTERNATIONALITY AND INTERDISCIPLINARITY

M. REUBER, F. KLOCKE

# NEW DEMANDS ON ENGINEERS

*Paths in Education leading to Professional Qualifications*

**Abstract.** The image of a career in engineering is changing. The climate of change in the industrial environment means that the engineering profession is faced with the need to meet new and evolving requirements. It is essential that the new generation of engineers, is qualified accordingly. The modified and extended profile of requirements also makes demands on the education and training programs for engineers. The form and contents of all education and training courses must be scrutinised and adapted continuously in order to ensure that engineering students are, and continue to receive optimum education and training for the functions which they will be required to fulfil in science and in industry. The following article examines the requirements to be met by the engineering profession and describes structures and approaches currently pursued in engineering faculties. In conclusion, suggestions are made with regard to the organisation of future-oriented university education in engineering.

## 1. INTRODUCTION

We are surrounded by products which are either used in technical applications or which were manufactured in technological operations. These range from domestic appliances through the majority of engineering tools used in our daily work, to virtually all modes of transport. Engineers are involved in the development and production of all of these technical goods. The function of the engineer, is to apply scientific knowledge and mathematical techniques in order to manufacture technical products.

The objective of engineering activity is to enhance the opportunities open to humans, by developing technological resources and by making intelligent use of them. The fact that attitudes in Germany towards engineering, have remained very positive over the past 20 years, bear witness to the success of this principle. Consequently, the engineering profession is held in high regard in German society and is regarded as competent and rational. Despite this generally very positive image, the engineering profession is currently the subject of lively discussion focussing on one hand on the concern surrounding the current lack of engineering graduates emerging from universities and training institutes and on the other hand, on the continuous and rapid changes in the demands which the engineering profession is expected to cope with.

The points described above result inevitably in new demands with regard to the education and training of engineers. Degree courses should be made more attractive, as a means of encouraging greater numbers of young people to study engineering and they should be shorter and more flexible, in order to provide the labour market with engineering graduates more quickly than is currently the case. The degree courses should also be adapted to suit the new shape of engineering, in order to give the students adequate preparation for the changing requirements of their chosen working environment, Figure 1.

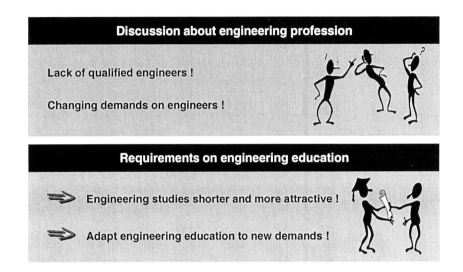

*Figure 1. Engineering Profession and Education.*

## 2. REQUIREMENTS TO BE MET BY MECHANICAL ENGINEERS

Engineering is a very varied profession and provides opportunities for activities in research and development, design, planning and project management, manufacture and assembly and in the sale and after-sales service of engineering products. Because of these very varied areas of application, industry demands that engineers should have an extensive range of abilities and skills identified in a series of surveys and studies, Figure 2.

The core area of expertise of engineers, is still their technical qualifications, i.e. they must have a firm grasp of the mathematical, scientific and technical basics. As a rule, they have thorough and extensive knowledge of one specialist area. Their qualifications extend beyond basic knowledge of their specialist subject to encompass the ability to apply the knowledge acquired, knowledge of the methodology of engineering work and the ability to think in systems, which they use to integrate their own work and developments within an overall concept and to evaluate it.

Engineers should also have knowledge extending beyond the boundaries of their own areas of technical expertise, i.e. they should be just as conversant with the principles of industrial management as with those of project and time management. Foreign language skills and basic knowledge of information and communications technology are also required.

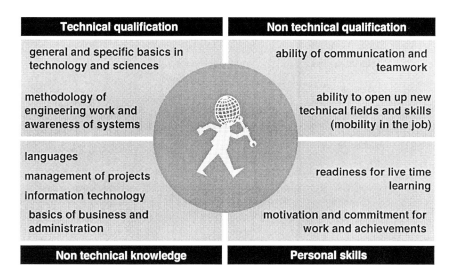

*Figure 2. Engineering Profile.*

Skills extending beyond boundaries of the technical focus concerned, are the characteristics which permit engineers to act effectively within the industrial and non-industrial environment. These include communication skills and the ability to work in a team or the ability and willingness to familiarise themselves with new specialist areas. Only those engineers who cultivate these skills can ensure continuing professional mobility, despite high levels of specialism.

Commitment and the desire to perform well, the ability and willingness to engage in life-long learning, creativity, flexibility, mobility and tolerance are the most sought-after and outstanding personal characteristics.

The scope of characteristics not directly connected with the technical knowledge and qualifications demanded in addition to the technical qualifications, is striking in the diagram in Figure 3. Although technical skills continue to be the basic qualifications of engineers, these same skills are considered by industrial organisations to account for only about 50 % of the ability to work effectively in an engineering environment. Personal characteristics contribute around a quarter, non technical qualifications and knowledge each contribute approximately 15 % to the professional qualifications of the engineer.

All this should not encourage people to conclude wrongly that specialist qualifications can be ignored. They continue to be the prerequisite for any type of engineering activity, but must be supplemented by knowledge and skills which go beyond the boundaries of the specialism concerned.

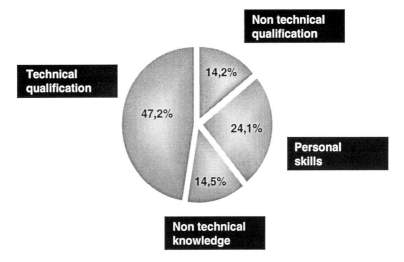

*Figure 3. Demands on Engineering Qualification.*

## 3. CHANGING IMAGE OF THE PROFESSION

The question as to why the demand for non-engineering key qualifications is increasing, is an interesting one. Some examples of the changes currently taking place in industry and their effects on the demands consequently imposed on the engineering profession are outlined and discussed in the following, Figure 4.

### 3.1. Products are changing

The complexity of technical products is increasing and their functionality is expanding. There are hardly any of the purely mechanical product functions which once made up the classical area of expertise of the mechanical engineer – modern products are units combining mechanics, electronics and software.

Engineers are compelled to have more interdisciplinary skills and must therefore be given the opportunity to absorb other, related disciplines in the course of their education and training program. Interdisciplinary thinking and the ability to cooperate in interdisciplinary team work are key factors, particularly in the area of product development.

### 3.2. Production is changing

Product development times are becoming shorter and innovation cycles faster. This demands greater cooperation between different areas within an organisation and the break-down of work-related thinking and action into areas such as development, design and manufacture.

Areas in companies are moving closer together – making basic knowledge of the work in neighbouring departments essential. Project management skills are becoming more important and the formation of interdepartmental project teams is demanding increasing levels of teamwork and team leadership skills.

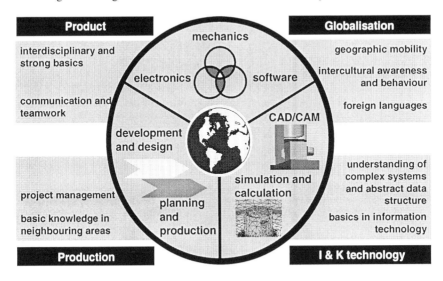

*Figure 4. Changes in Production Environment and Impact on Engineers.*

*3.3. Information and communications technology is altering working life*

The farthest reaching changes in working practices are associated with the introduction of information technology to virtually all areas of manufacturing technology. Information and data flows in companies are increasingly being digitised, starting from product data through to the logistics in production and sales as well as inventory management. New simulation techniques provide means of calculating machine behaviour, manufacturing processes and part strain, thus clearing the way for more efficient and more economical product development.

Engineers must have basic skills in information and communications technology and must be able to think in complex systems and data structures. They should also have firm knowledge of the potential of information technology be able to estimate the extent to which it can be applied.

*3.4. Companies are increasingly becoming globally active*

Globalisation is a further driving force behind change. Companies produce and are active worldwide. They frequently demand correspondingly high commitment to geographical mobility from their staff. Employees must also show the ability to interact competently with people from other cultures and a capacity for tolerance, which will enable them to work in international teams. They must develop the awareness that manners, decision-making and working practices are different in other cultures than in their own environment.

Naturally, this involves the acquisition of foreign language skills – English has become less of an additional qualification than a natural prerequisite for engineers.

## 4. A COMPARISON OF THE PATHS THROUGH EDUCATION AND TRAINING

Having addressed the qualification characteristics needed for a career in engineering in the previous section, the following sections focus on discussion of the various education and training routes pursued in different countries. The single-track model used in Germany is compared, for example, with the consecutive curriculum found predominantly in Anglo-Saxon countries.

### 4.1. Educating and training engineers in Germany

The education and training of engineers in Germany, is characterised by two parallel strands with different qualification profiles – a degree course at a university or institute of technology and the degree course at a technical college, Figure 5.

The classical difference between the two types of higher education is that a university education is more theoretical and more scientific whereas the education and training offered by the technical colleges is highly application and practice-oriented.

- **Different profiles and university degrees**

- **Emphasis on science and application**

- **Qualification beside university studies**

- **Degrees are clearly defined and protected by law**

⇨ Horizontal structure for a variety of tasks in industry and personal preference

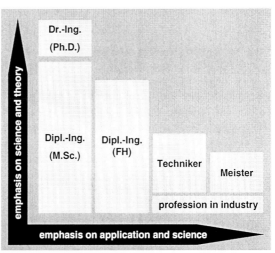

Figure 5. Technical Education in Germany.

This difference is no longer so clearly defined and is the subject of lively discussion. On one hand, many technical colleges would like to expand their profile to include more science and on the other hand, many universities are increasing the level of application and practice-orientation, particularly where a strong infrastructure of application-oriented contract research already exists.

Nevertheless, these profiles show the historical and differences, which on the whole have been retained and which are reflected in the differences in the nature of the education offered and degrees awarded. Figure 5 also shows some of the several paths for gaining technical qualifications beside the higher education such as apprenticeships and extended qualifications focusing strongly on shopfloor practice.

The education and training system provides various courses and degrees, which correspond to the personal preferences and talents of the students and which are recognised as serving the wide range of engineering activities and functions to be performed within companies.

### 4.2. The consecutive study structure

The most striking characteristic of the consecutive structure is the sequential sequence of degrees, each of which builds on the previous one. The Bachelor of Science is gained in the first course before further study can commence for the Master of Science course, which is comparable with University Diploma in Germany, Fig. 6.

The Bachelor (B.Sc.) degree is regarded as qualifying the holder for employment as an engineer and is achieved in 3-4 years, depending on the country and university involved. This degree is the basis on which the vast majority of students will seek employment. The M.Sc. can likewise lead to employment in industry but it is also the prerequisite for a farther-reaching scientific career at a university.

**Characteristics**

- Sequence of subsequent studies and degrees
- First degree (qualification for a job in industry) is given after 3 - 4 years
- Entrance exams for university and graduate studies

**Heterogeneous educational offers**

- Specific and diverse solutions in educational offers ranging from application oriented to scientific orientation
- Quality and acknowledgement of degree depends strongly on institution and studies

*Figure 6. Consecutive Studies.*

As a rule, the applicants for degree courses sit entrance examinations before beginning to study and between courses. The results of these examinations in conjunction with the results already achieved, are used to decide whether applicants will be admitted to the course of their choice. In some cases, they are accepted for courses with the proviso that certain conditions must be met.

The education systems which follow the consecutive structure are not nearly as homogeneous as the basic principle of the study structure outlined here, suggests. They provide a multitude of individual solutions, which differ from country to country and which vary considerably in quality and orientation even within the individual countries. Here too, both practical, career-specific courses and scientific, general courses are offered. In other words, there are differences between university degrees, although there is no formal, institutionalised differentiation between education and training profiles such as the distinction in Germany between FH (Technical College) and TH (Institute of Technology).

As a result of the wide range of university degrees awarded, the level of recognition depends largely on the reputation of the institution attended and on that of the course which has been completed.

*4.3. Internationalisation of teaching and learning*

The globalisation of the markets is also influencing the nature of courses offered and is promoting the effort to ensure that the education and training of engineers becomes more international. Periods of time spent by students in other countries improve their knowledge of foreign languages, encourage global transfer of knowledge, and enable students to acquire inter-cultural skills. However, if a global exchange of students is to become reality, it is essential to organise internationally compatible degree courses which will permit students to move effortlessly between courses of education and training in various countries, Figure 7.

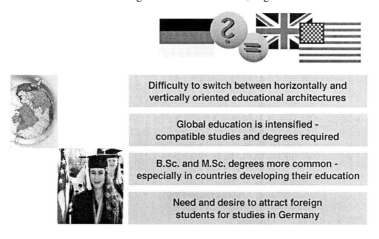

*Figure 7. Comparison of Approaches.*

Because of the structural differences outlined above, this transparency between the horizontal, single-track structure in Germany and the vertical division of the consecutive courses, is patchy. The cross-over to a defined point in the diploma course is difficult, particularly for those who have a B.Sc. as a first, vocational degree when they come to Germany. Possible changes to the degree course structure in Germany are therefore the subject of discussion with a view to achieving greater transparency between German courses and the internationally very widely practised and recognised Bachelor and Master degrees.

The aim is to encourage foreign students particularly from the emerging economies and from developing countries, to study in Germany. This will ensure the long term competitive strength of German industry in the environment of global trade, since those who study in Germany today, will later become mediators and contacts for German industrial organisations and key personnel in opening up these markets. Germany is competing with industrial countries, which have consecutive study structures much more like those in the emerging and developing countries, in this vital struggle to woo foreign elites.

## 5. MEASURES TO REFORM UNIVERSITY ENGINEERING COURSES

Two major demands emerge from the previous discussion of the issues concerned. First is an increased requirement for qualifications as a result of changes affecting the profession, second are the strong efforts to provide an internationally oriented degree course which is capable of fulfilling the requirements of a global market in education and training. Measures aimed at increasing the level of qualifications focus mainly on teaching skills and knowledge about non-engineering subjects and on increasing the capacity for interdisciplinary thinking, Figure 8.

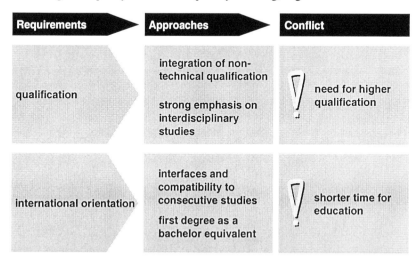

*Figure 8. Requirements and Approaches on Study Reforms.*

The prerequisite for internationalisation of the degree courses is compatibility with the consecutive model. It is in this context that the frequently expressed demand for a vocational short degree course corresponding approximately to the level of a Bachelor degree and which will provide a vocational qualification after 3-4 years.

At this point, it becomes clear that there is a fundamental conflict of goals, which is very rarely referred to directly in the discussions revolving around study reforms: an increased requirement for qualifications is not necessarily compatible with the demand for shorter study times, Figure 8.

*5.1. Case study – Model course in production technology*

Approaches and measures, which are capable in principle of achieving both objectives, are outlined in the following on the basis of a case study. These measures may not be capable of really reconciling the conflicting aims but they are all the more important since their systematic implementation will help to bring about a compromise solution.

The approach starts from the maximum demand, for a vocational degree which should be achievable – as previously outlined – after a study period of approx. 3 years. If a high level of vocational skill to be achieved from the beginning, it is vital to ensure that the course focuses from the early stages, on one area of application. A basic course in production technology is therefore outlined in the following as an example. The fundamental orientation towards production technology as an area of application must not be regarded as any kind of early specialisation since the aim is to ensure that professional mobility is not restricted. The aim is, rather, to illustrate the application of knowledge on the basis of problems occurring in the chosen area of emphasis, Figure 9.

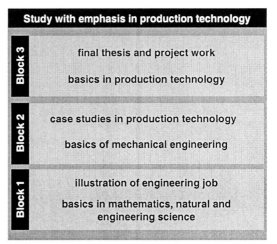

*Figure 9. Inherent Measures for Qualification.*

The intention behind the concept is to foster general technical understanding and interdisciplinary basic knowledge and, at the same time, to promote the occupational skills of the engineers after the degree, by confronting them with numerous practical problems and by providing clear examples. Graduates of university courses of this nature have a fundamental knowledge of engineering science, of mechanical and production engineering. They have problem-solving and transfer expertise, are familiar with the techniques required in engineering and have basic knowledge of scientific work. From this point of view, they are particularly useful in the field of production technology but are professionally mobile, due to their basic knowledge and expertise in applying engineering techniques. Within the course, there is gradual consolidation of the subject matter in three stages, allied with more detailed study. Traditionally, general engineering principles are studied before fundamental principles of mechanical engineering and then fundamental principles of production technology are introduced, Figure 9.

The specialist training should be accompanied by a series of measures aimed at increasing the motivation of the students and at fostering the achievement of qualifications not directly related to the engineering specialism.

The early study phases focus on two aspects, Figure 10. First, fostering an understanding of the engineering profession and of the multi-faceted areas of activity for engineers. Second, increasing awareness of and capacity for interdisciplinary problem-solving.

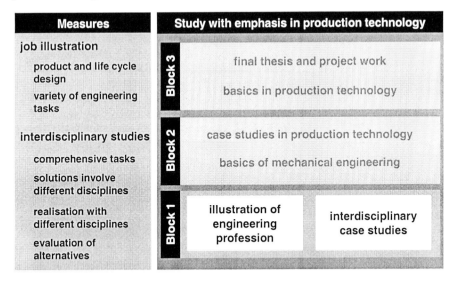

*Figure 10. Beginning and Orientation.*

The students should be familiarised as early as possible with a comprehensive view of the engineering profession. This can be achieved by studying the product development process or – even better – a product life cycle, which lends itself to a full description of the principal areas of work for engineers. A course of this nature, called "An Introduction to Mechanical Engineering" is offered at Aachen University of Technology, for example.

It is conceivable that interdisciplinary case studies, involving the study of entire problem areas and requiring either the participation of students from various disciplines or in which there are various problem-solving options from different disciplines, could be introduced as a means of increasing interdisciplinary awareness. It is possible that a functionality could be presented, which can be achieved either electrically or mechanically. This provides a starting point for the understanding of the way in which mechanics and electronics can work together and for the art of assessing various solutions and of discussing advantages and disadvantages.

Non-specialist knowledge should be integrated in the second study phase in addition to the principles of mechanical engineering, Figure 11. The theoretical principles studied in the specialist phase should be illustrated at this point largely via practice-oriented problems from the consolidation phase. However, this requires greater diversification of example applications in practical courses, corresponding to the different areas of consolidation and intensification. Accordingly, more emphasis must be placed on teaching the basic subjects.

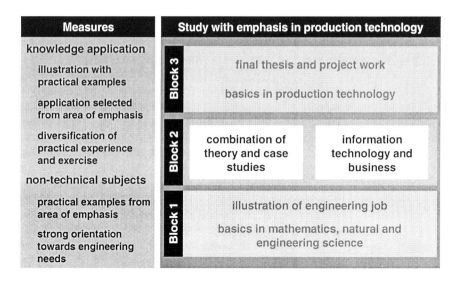

*Figure 11. Intermediate Stage with Basics in Mechanical Engineering.*

Cross-curricular knowledge of business management and information technology should be disseminated, as far as possible, via subject-specific consolidation problems. In each case, it is vital to ensure that the curriculum is tailored to the needs of engineers.

The subject-specific contents of the area of emphasis, which should be supported to a considerable extent by case studies taken from the field of production technology in order to ensure that subject-specific problem-solving skills are developed, are studied in the last phase, Figure 12. In this phase, the contents are so focussed on theoretical principles and on problems arising in industrial practice, that theoretical principles and issues related to industrial practice, can easily be combined. Students can select their additional main subject area of interest within production technology, via electives and research subjects.

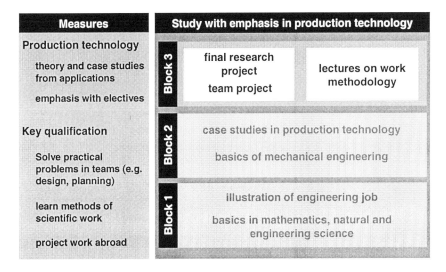

*Figure 12. Final Stage with Emphasis in Production Technology.*

The cross-curricular link subjects are given additional sustained emphasis by the project work and final exams in the concluding phase of the degree course before commencing joining the profession.

Solutions to practice-oriented problems should be developed in team-work as part of the project work. Additionally, a concluding scientific essay should be set and accompanied by the teaching of scientific methods. A concluding piece of work of this nature also offers an excellent opportunity for a prolonged period abroad.

*5.2. Analysis and evaluation of the study model*

The orientation of the study profile outlined here, prepares students both for activities within industry and for the achievement of further scientific university qualifications. The aim is to provide education and training with a scientific basis and with a strong practice-oriented slant, which ensures a high level of professional skill from the outset.

There are a number of unanswered questions at this point. The most important of these is whether a university course of this nature is actually capable of delivering the depth of learning which is both desirable and vital within only three years. Can the graduates acquire a sufficiently high level of professional skill to gain acceptance in industry? The answers to these questions depends largely on the regard in which the subject qualifications of the graduates of courses like this, are held. It is certain that the contents of the diploma degree in Germany will have to be reduced considerably if this substantial reduction in the duration of studies is implemented. It is difficult to predict how the labour market will react, particularly as the technical qualifications of engineering graduates are currently regarded as virtually ideal.

It is generally agreed that 5 years of study at university certainly provide a suitable time-frame in which to ensure thorough education and training of engineers equipped to face the future, predominantly via the measures already in place. Nevertheless, in the long term, engineering faculties in Germany will not be able and willing to avoid the pressure for international orientation of education – particularly in view of the need to make our education attractive to foreign elite groups.

In conclusion, it can be said that the conflict of aims between continuously expanding curricula on one hand and demands for shorter study duration on the other hand, cannot easily be solved and still provides plenty of subject matter for discussion. A number of approaches to this issue have already been developed. Many of the models go far beyond the measures discussed here. However, the fundamental fact remains, that massive efforts are needed both now and in the future, to find suitable and satisfactory compromise solutions to the multi-facetted and in some cases, diametrically opposed demands being made on the education and training of engineers. Prerequisite for a future-oriented education is that above considerations are taken into account in all dicussions on form and content of education and training courses. Being aware of the conflict of aims will help to define goals and find appropriate approaches for continuously adapting education and training to new and changing demands on the engineering profession.

## 5. SUMMARY AND OUTLOOK

The career as an engineer is particularly multi-facetted and offers excellent prospects for young people with an interest in technology. The education and training system currently in place in Germany for engineers, is acclaimed worldwide in terms of engineering excellence and offers a wide range of horizontal opportunities to acquire qualifications. Its weaknesses are perceived to be the lack of opportunity to acquire non-engineering-specific expertise and in the lack of compatibility with the consecutive curriculum, which is offered in many other countries.

The principal aims behind the efforts to bring about reform, are to ensure future-oriented, flexible education and training and international compatibility of the education structure. There is considerable potential to optimise the education and training of engineers, by using existing structures systematically in order to provide opportunities for the acquisition of non-engineering-specific key qualifications. One of the main positive aspects of this approach, is the high didactic value which results from the shared teaching of specialist and personal skills.

## 6. REFERENCES

Doege, E.: 2000 „WGP Standpunkt - Besonderheiten der deutschen Universitätsausbildung", *wt - Werkstattstechnik,* Vol. 90 (2000), p. 274

Eversheim, W., M. Erb: 1997, „Gedanken zur zeitgemäßen Ausbildung von Studierenden", *wt - Produktion und Management,* Vol. 87 (1997), pp. 18-20

Franz, H.: 1997 „Der Ingenieur in der Industrie", *wt - Produktion und Management,* Vol. 87 (1997), pp. 9-12

Henning, K., J.E. Staufenbiel: *Das Ingenieurstudium – Berufsziele, Studieninhalte und die Wahl der Hochschule.* Staufenbiel Institut für Studien- und Berufsplanung, 6. Auflage, 1999

HIS: *Ausbildung und Qualifikation von Ingenieuren – Herausforderungen und Lösungen aus transatlantischer Perspektive.* HIS Kurzinformation A6/98, Hannover, Germany 1998

IW: *Der Ingenieurberuf in Zukunft: Qualifikationsanforderungen und Beschäftigungsaussichten.* Impuls-Stiftung/Institut der deutschen Wirtschaft Köln, Köln, Germany 1999

Keedy, L.: *In Stufen zum Ziel – Zur Einführung von Bachelor- und Master-Graden an deutschen Universitäten.* Raabe Verlag, 1999

Minks, K.-H., C. Heine, and K. Lewin: *Ingenieurstudium – Daten, Fakten, Meinungen.* HIS Hochschul-Informations-System, Hannover, Germany, 1997

N.N.: *Bachelor und Master in den Ingenieurwissenschaften.* Tagungsbericht DAAD and HRK, DAAD Dokumente und Materialien Vol. 32, Conference May 25/26 1998, Bonn Germany, 1998

Staufenbiel R.W.: 1993 „German Education in Mechanical Engineering, from The Perspective of the RWTH Aachen ", *Int. J. Engng. Ed.,* Vol. 9, No. 1, pp. 29-42

Taurit, R.: 1993 „German Engineering Education from a Fachhochschule Perspective", *Int. J. Engng. Ed.,* Vol. 9, No. 1, pp. 20-28

Treichel, D.: 1997, „Interdisziplinäres Ingenieurstudium", *wt - Produktion und Management,* Vol. 87 (1997), pp. 29-32

VDI: *Empfehlung des VDI zur Integration fachübergreifender Studieninhalte in das Ingenieurstudium.* Verein Deutscher Ingenieure VDI, Düsseldorf, Germany 1997

VDI: *Empfehlung des VDI zur Integration fachübergreifender Studieninhalte in das Ingenieurstudium.* Verein Deutscher Ingenieure VDI, Düsseldorf, Germany 1997

VDI: *Ingenieur – Berufsbild im Wandel.* VDI Nachrichten Fazit, Düsseldorf, Germany 1997

VDI: *Ingenieurbedarf 2000 – Eine Studie des VDI..* VDI Nachrichten Fazit, Düsseldorf, Germany, 2000, http://www.vdi.de

VDI: *Memorandum des VDI „Zum Wandel des Ingenieurberufsbilds".* Verein Deutscher Ingenieure VDI, Düsseldorf, Germany 1997

VDMA: *Internationalisierung der Ingenieurausbildung – Die neue Herausforderung für Hochschulen in Deutschland.* Empfehlungen von VDMA und ZVEI, Frankfurt, Germany, 1997

VDMA: *VDMA Ingenieurerhebung 1998.* Verein Deutscher Maschinen- und Anlagenbau VDMA, Frankfurt, Germany, 2000, http://www.vdma.de

Zwick, M.M., O. Renn: *Die Attraktivität von technischen und ingenieurwissenschaftlichen Fächern bei der Studien- und Berufswahl junger Frauen und Männer.* Edited by Akademie für Technikfolgenabschätzung Baden-Württemberg, Stuttgart, Germany, 2000

*Martin Reuber, Fritz Klocke*
*Aachen University of Technology*
*Aachen, Germany*

P. BASCHERA, N. RICKERT

# DEVELOPING LEADERS FOR TECHNOLOGY INTENSIVE COMPANIES

*The UNITECH International Story*

**Abstract.** The Hilti Corporation, headquartered in the Principality of Liechtenstein, holds a leadership position world wide in the specialised field of fastening and demolition systems for the construction industry. As in industry generally, Hilti has difficulty in finding highly qualified engineers who are able to move at ease in international environments and have a feeling for the management side of the business as well. To address this gap, Hilti and the ETH Zurich initiated the UNITECH International Society. Today, it comprises seven leading European technical universities and twenty leading multinationals. Together, they set up a programme which provides excellent engineering students with an opportunity to complement their studies with management courses, cross-cultural experience and language skills (exchange term) as well as industry and management exposure (management internship). Through group work and exposure to industry, these students are also able to build up interpersonal skills which are a crucial element in today's business life.
In its first sections, the following article describes how UNITECH developed from the initial idea to a multinational network of leading technical universities and multinational corporations. The second part describes the programme, explores the present culture of the organisation and takes a look into the future.

## 1. OUTSET SITUATION

Hilti - as in industry generally - has been exposed to major changes in its environment. These trends, such as globalisation, technological revolution and the tremendous developments in information technology, have not only impacted the way Hilti does business, but have also changed the requirements that its employees have to meet.

Where engineers are concerned, it has become more and more difficult to find employees who meet these changing requirements. Today's engineers need to be "knowledge workers", "culture surfers" and strong "networkers" to use catch words. While today's educational systems is producing excellent specialised engineers with strong technical Know-how, these people often lack the cross-cultural experience, language skills, interpersonal skills and an understanding of management principles that would make them ready to cope with the above mentioned trends.

## 2. DEVELOPMENT OF UNITECH INTERNATIONAL - THE STORY

The idea of addressing this gap in university education by founding a joint initiative of industry and academia came from the Hilti Corporation mid 1998. At that time, Hilti had already gained extensive experience through co-operation in the field of academic education, e.g. playing a key role in the CEMS (Community of European Management Schools) network. Today, both organisations are clearly different, but co-operation has already begun as they share a similar spirit and their student target groups complement each other. From these experiences we learned a lot for the build up of UNITECH International.

## 2.1. Building the basis

Through existing links among senior Hilti and ETH Zurich executives, the feasibility of the idea and the interest on the academic side were explored. This approach was key to the success for two reasons: the choice of the right partner (ETH Zurich) and the top-level commitment. Without these two factors, it would probably not have been possible to push this project ahead at this speed and to bring the right group of academic and corporate partners together with such ease.

On the academic side, the Rector and the Vice Rector of International Affairs of the ETH were able to build on their strong international contacts, e.g. the IDEA league or the CESAR network, to identify and approach the right partners. At the same time, Hilti used its top executive contacts, on a CEO and Board level, with selected industry leaders to establish the basis for the corporate partner group. This resulted in a first group of around 10 corporate partners with roots in the German speaking area.

Once the academic partners had been united, each university approached Industry partners in its own environment in order to achieve a good cultural mix on the corporate side. The fact that all these contacts were made at such a senior level (rector, vice-rector, CEOs or board members) ensured that the programme was well positioned in the participating institutions. As a result, UNITECH was perceived as a joint network of strongly involved stakeholders instead of a sponsoring partnership right from the beginning. This was especially important on the industry side. Many activities of industry in academic environments are focused very much on recruitment. Such a narrow focus, however, limits the potential of co-operation as it highlights the competitive element between the various corporate partners while neglecting their potential joint interests. In this first phase of the project, the development work was done in a relatively small team comprising only the ETH Zurich and Hilti representatives.

## 2.2. Identifying and involving potential partners

In the second half of 1999, about a year after the initial discussions between Hilti and the ETH Zurich, the second phase of meetings started. In a first step, other academic partners joined in the discussions (RWTH Aachen, UPC Barcelona, TU Delft, Imperial College London, Politecnico di Milano). Although many questions were still open, this group took the very courageous decision in the autumn 1999 of already starting the programme in the following autumn. Taking this decision at a moment, when much of the programme was still unknown and only few industry partners were firmly committed, was probably the key breakthrough in this development. It is doubtful that the programme would exist today if this decision had been postponed because much of the dynamics would have been lost by planning the first exchange for September 2001 instead of for September 2000. Once the go-ahead had been decided, the search for further corporate partners was intensified and, in February 2000, the first corporate partners meeting took place with 17 interested parties represented.

*2.3. Building understanding and a consensus between academia and industry*

In the first half of 2000, meetings of the corporate and academic partners were still held separately. During the academic meetings, Hilti took part to bring in the corporate view and, during the corporate partners meeting, ETH representatives spoke on behalf of the academic partners. Although both the ETH and Hilti had no official mandate from their respective groups, they naturally held special positions due to their previous involvement and their resulting expertise in UNITECH matters.

Looking back, it was right to split the groups at the beginning of the second phase of the project. It allowed each group to focus on their specific issues first, e.g. on questions of academic standards, e.g. credits, and on the operations of such a programme among academicians as well as questions of internships, finances and organisational issues among corporate partners. Separate meetings, however, had several disadvantages. For instance, it made it difficult for existing stereotyped views about the other party to be overcome. While corporate partners perceived academic institutions as slow and conservative, the academicians had a strong fear that the corporate partners would want to interfere with academic freedom. It also became more and more difficult for Hilti and the ETH Zurich to represent the point of view of their respective groups as, needless to say, the interests and opinions differed within each of them.

Both corporate and academic partners thus expressed the wish that a joint meeting should be held. It then took place in June 2000. It was another key event in the entire development phase as it resulted in a major boost in mutual trust. Both sides could feel that their partners had a genuine interest and a strong commitment to this project. Today, it is still UNITECH's philosophy to foster live interaction among all UNITECH partners. Although it is a time-consuming and costly way of running such a network, it is probably the only way to establish the level of trust necessary for developments to go beyond a sponsoring partnership.

*2.4. Founding of the legal organisation and kick-off of the programme*

The last step of the initial development of the UNITECH network started on September 1, 2000 when the UNITECH International Society was founded at Como in Italy. The event was embedded in the so-called "start-up" week of the first academic year and it thus included the first UNITECH students. The list of participants, comprising several rectors and deans as well as a large number of corporate senior executives, was a clear sign of the importance attached to this network by the partners.

## 3. THE RESULT: THE UNITECH INTERNATIONAL PROGRAMME

The aim of the UNITECH International Programme is to provide outstanding engineering students with an opportunity to complement their studies with management courses, cross-cultural experience and language skills as well as industry and management exposure. Through this exposure they develop the crucial interpersonal skills for today's business life. In this section the elements of the programme are briefly described without going into the technical details of credit requirements.

### 3.1. Academic exchange

UNITECH students study at least one term at one of the partner universities. The chosen university must be in a different cultural and language area, thus strictly excluding exchanges between Aachen and Zurich. During this exchange period students follow local courses in management and engineering topics. Some universities have started to offer selected courses exclusively for UNITECH students. The majority of courses are however taken together with local students ensuring a strong cultural exposure. By choosing between one or two academic terms abroad, the student can decide if he or she wants to focus more on the academic or the industry part during the UNITECH year.

### 3.2. Internship in industry

The UNITECH internship is not a technical placement. Corporate Partners have committed themselves to provide students with internships which provide them with an insight into aspects of corporate management in practice. For many corporate partners this commitment means that they have to create special placements, which go beyond their regular internship programmes. Hilti has for example chosen a two-fold approach. It offers UNITECH students projects at the interface of management and technology with a final presentation to senior management. On top the students will participate in an Executive Board meeting and join the Executive Board on a visit to a Marketing Organisation. Thus students are given insight into all levels of management decision making.

The internships have been the major focus of the Corporate Partners Committee, being addressed at every meeting to share different approaches and experiences. At present, students are required to work at least three months (or six months if they study only one term at a partner university) with a corporate partner.

### 3.3. Joint Modules

A unique feature of the programme are the so-called "Joint Modules". These are courses, where all UNITECH students come together. In the academic year 2001/2002 three such Joint Modules will take place.

The "Start-Up Week" is the kick-off for the programme. It takes place at the very beginning of the academic exchange. The focus is on basics of management and the week is built around a business game, a non-computer based simulation, which is developed specially for the UNITECH needs. The game is complemented with lectures and workshops of academic and corporate lectures, providing the theoretical background for the gaming experience. Interpersonal skills are developed through group interaction, cross-culture and project management workshops and discussions with senior industry and university representatives. An essential side-effect of the event is a strong network-building between the students. As the event overlaps with the "End-of-year" event for the students of the previous academic year and the General Assembly with corporate and academic partners the networking is extended to all UNITECH stakeholders.

The "Mid-Term week" takes place three months later, when the students are in the middle of their academic exchange. Besides further strengthening of the group feeling and networking - generally referred to at as the "UNITECH spirit" - it provides students with further input from academics and industry representatives. The focus of this week is on more strategic issues and students learn to work based on case-studies. In addition they will write papers in groups, thus also providing an own input to future UNITECH teaching activities.

The exchange year is completed by the "End-of-year Event". Besides an additional investment in interpersonal skills the focus is once again on strengthening networking among students as well as between the students and the other stakeholders. The event overlaps with the Start-Up week of the next academic year and the General Assembly. Students, respectively alumni – as part of their responsibilities in the programme – act therefore also as mentors for the following students, as they have gained valuable experience about studying and working abroad.

These Joint Modules are by far the largest position in the UNITECH budget, which is financed by the corporate partners. Through their investment decisions, e.g. to add the "Mid-Term week" to the programme, the Council made a clear statement that it considers joint events a central element to differentiate the UNITECH programme from other exchange schemes. As all events are joint productions of the academic and corporate partners, they also foster cooperation within the network.

*3.4. UNITECH degree*

UNITECH students complete their programme when they have fulfilled all the UNITECH requirements and finished their home-degree. From that moment on they automatically become UNITECH Alumni. In the face of present developments concerning accreditation of degrees such as the Bologna declaration and the bachelor/masters discussion in general, the creation of a formal degree has not been given first priority. The understanding is that the important outcome of having completed the UNITECH programme is a better qualification for careers in industry and not a formally accredited degree.

Major investments and efforts are however taken in order to quickly achieve a high brand recognition of the "UNITECH-degree", especially among corporate and academic partners. Considering the fact that corporate partners represent several million employees around the globe, achieving this has a much greater effect than formal issues.

*3.5. Alumni Organisation*

Already in the very first meetings, ETH Zurich emphasized the importance of a well-functioning Alumni Organization. The Alumni Organization has several functions. First of all it is a means to foster continuing learning and networking among the UNITECH graduates. UNITECH Alumni should continue to meet on a regular basis, e.g. at seminars or social events. The Alumni are however also considered as one of the stakeholders of the programme. They will be represented in the Council and contribute with their experience and their resources to the development of the programme. It is planned to establish the Alumni Organisation at the second General Assembly in September 2001, when the first students complete their UNITECH year. The students have set up a work group which has been charged to develop statutes, therefore taking responsibility for their own organisation.

## 4. TODAY'S STRUCTURE AND CULTURE

Although a suitable organisational structure is not a guarantee for success and the desired culture, it is certainly an important factor. In view of this, the structure and decision making of the UNITECH International Society have been topics of many discussions and meetings. The objective was to create a structure which involves all stakeholder groups in decision making, but remains flexible and agile for response to new needs.

The approach chosen to achieve this was to have a joint strategic body, comprising representatives of academic partners, corporate partners, students and, in the future, alumni too. As a result of the experience of its members - most are senior executives of the academic and corporate partners - this body is the strategic backbone of the organisation. Apart from the council, three other committees exist, each with a decision-making function in its own field and an advisory function for other subjects. They are the Academic Committee (responsible for the curriculum), Corporate Partners Committee (responsible for internships and corporate partner input for the programme) and the Local Office Committee (responsible for operations of the programme). The committees meet about three times a year. An International Office in Zurich supports these committees in their work and co-ordinates all UNITECH activities.

It has become customary for representatives of other committees to join in meetings to ensure that there is a good flow of information and expertise. Although the relatively large number of meetings is a costly and time-consuming way of maintaining and developing a network, it is probably the only way for an organisation with UNITECH's aims and philosophy.

The most important achievement in a cultural respect is the level of trust which has been established among the partners. This is also important within each stakeholder group. The academic partners have, for instance, gained a good understanding of each other's situation. This should not be underestimated as harmonisation of university education has not yet progressed very far, and this has led to very specific local environments. Understanding of this kind helps to develop new elements in the curriculum and opens doors to other activities not directly linked to the core programme, such as co-operation in the field of e-learning and other areas.

The same goes for the corporate partner group. Each corporate partner can veto the application for membership from a direct competitor. Although no use has yet been made of this right, it has resulted in efforts to have a wide range of industries represented rather than several partners from the same business sector. Consequently, there are very few companies in strong business competition with each other that also foster mutual trust in one and the same group. A, so to speak, "code of conduct" supports this trust which is not yet in written form, but emerges as a pattern of decisions. These are mainly decisions when it comes to contact with students. So far, the decision of the corporate partners has been that they compete for students through their active participation (internships, lectures, etc.) in the programme and not through financial investments and incentives. Recruiting activities, for instance, are kept low key without fancy advertising or event sponsoring. This common understanding creates a good basis for fruitful networking among the partners, which has already led to many formal and informal meetings on joint issues, e.g. benchmarking and exchanges of experience.

But not only trust within one group of stakeholders, e.g. between corporate partners, is important. Most additions to the programme involve both sides, academia and industry. Therefore each side has to trust the other that they are both capable and willing to take their part of the responsibility and actions. A good example is the student selection process. It takes place in two steps: an academic selection with focus on academic excellence, executed by the partner universities and a selection done by corporate partners which emphasis people and leadership skills. Here again, trust in decisions reached by the other partners is a must.

## 5. OUTLOOK

At present, UNITECH has seven academic and twenty corporate members. Sixty students have been selected for the academic year 2001/2002. The target for the next couple of years is to enlarge the academic network by about two European partners and to increase the number of corporate partners to around 30. Given the present structure of the programme, this would lead to some 100 to 140 students a year, which is a relatively small number for such a network.

At the same time, however, the intensive cooperation as the key strength of the UNITECH International Society is also its limitation. A high level of co-operation and trust can only be maintained up to certain number of partners. From a geographical point of view, it would also be very difficult to integrate overseas partners if face-to-face interaction is considered key to UNITECH's culture. Today, this may seem a rather conservative approach, but, to date, it has kept strategic decision making successful. As a result, UNITECH will not go for size, but focus on quality and the intensity of relationships for the time being. UNITECH is an elite approach to develop future leaders for international companies with strong roots in technology.

As a consequence, the number of students per corporate partner will remain relatively low at three students per partner as a realistic figure. If the focus of partners was merely on recruitment or on fostering an academic exchange, the time and money involved would not be justified. UNITECH must thus focus on providing its members with added value beyond the core activities. Some has already emerged, e.g. strong and valuable networking. Other added value, such as joint projects in e-learning, are presently being evaluated. Further perspectives are that UNITECH transfers the expertise it develops when training and educating students into management development products for its corporate partners. These and other activities will add up to positive pay-back for the considerable investments made.

## 6. KEY LEARNING

From the initial idea to actual implementation of the programme (first students abroad!), it took only two years. This is a very competitive result and proves that academia - contrary to what is general thought – is able to react quickly to new needs. There are, however, some prerequisites:
- full commitment from all partners
- willingness to take decisions in uncertain context (go / no-go decision)
- familiarisation with each other
- trust
- open-mindedness

Although UNITECH has been a success story up until now, many things could have been approached in a better way. This is not surprising as, to date, relatively few reference models existed from which we could draw. Here some of our learning:
- There was late focus on installing an international office. The resulting lack of resources led to a temporary slow down of developments.
- The go-ahead decision reached at an early stage to get the program of ground was positive, but it also led to pending issues being postponed. These issues should have been made more transparent to avoid some "surprises" during critical stages of the project.
- Not enough focus was put on communicating programme such as curriculum requirements and internship criteria, leading to some inconsistency in their application.

UNITECH International clearly proved that projects like this go beyond a plain cost-benefit analysis, although these have also been carried out, and they are built on visions, passion and committed people that make it happen.

*Pius Baschera - Hilti Corporation, Schaan, Principality of Liechtenstein
and Swiss Federal Institute of Technology, Zurich, Switzerland
Nils Rickert - UNITECH International Society, Zurich, Switzerland*

M. POLKE

# CHALLENGES OF A VIRTUAL UNIVERSITY CAMPUS: UNIVERSITY POLICY AS A CONSISTENT SYSTEM

**Abstract.** The paper starts with a short historical review. Subsequently the relationship between New Technologies and structural changes in society and economy will be presented. Furthermore, the structures of information will be shown both through an object-oriented view and as flow-oriented systems and their elements. The roles of engineers will be described within this technological revolution. Through comparing the international Engineers Studies, it will be evident that only a consecutive model of Higher Education studies can be the goal of further university development offering the same starting conditions for everybody (e.g. after 12 years of secondary schooling). Thus, system analysis will be used to draft a *Virtual Campus*. It is suggested as one necessary step of further university development. Subsequently new university workflow and organisation structures are to be developed including new methods of teaching. The tools of operating cost analysis and the financing methods for the "new" Campus University are also to be adapted to the new goals. Finally training-methods and approaches to Further Education for qualified people need to be integrated into the Campus University consistently based on a demand analysis.

## 1. INTRODUCTION

Information has become the most important resource in all areas of our lives. It causes fundamental changes: our societies have become *information societies* (Polke, 1988; GKI, 1995). Hence Engineering education is changing as well, and we need to cooperate through exchanging our experiences within the global framework of Higher Education in order to cope with these challenges today. It includes to recognise the innovative structures within the *new developments* and to compare them with our *traditional structures*. Based on these views, the Working Group *Informatics and Computer Sciences* has been recently established. It comprises representatives from 12 different (German-speaking) professional associations on Computer Sciences and their applications. In 1997, the first outcomes of its work were presented to leading representatives of parliament and government, economy and university, professional associations and churches as well as schools and educational publishers. This first document deals with a new educational initiative in response to the fast changes of information technology today (GKI, 1997).

During the 18$^{th}$ and 19$^{th}$ centuries, it became a necessity for everybody within our societies to experience some schooling. Reading, writing, arithmetic were to be mastered by all people whether in administration or factory. Today it has become impossible for anybody to acquire all knowledge needed. Our information society appears *flooded* with information which claims to be necessary for everybody. Our existing educational system, however, appears to meet its limits as long as it tries to teach all such information within its traditional structures. It is definitely over-challenged with the impact of the new information and communication technologies - the New Media. Thus, we all are forced to continue learning beyond schooling. Lifelong learning may be considered *professional and cultural survival training* (Forum 2000, 1998).

There is the danger, however, that our societies may be split into two layers: those who master the challenge of the information society, and those who only marginally use the valuable information offered, for many reasons. Hence a thorough implementation of new media through all layers of society is needed. It is not sufficient merely to link the schools to the Web without fundamental changes in teaching and learning, neither is it sufficient to *invent* new buzz words in order to pretend that changes are taking place. On this background, a new expert Working Group has recently started to discuss the future of engineering education in Germany (KFM, 1999). The group had been invited by the VDE and VDI under the responsibility of the former VDE chairman, Hermann Wolters. In this report, some aspects of these discussions will be described leading to proposals which are to illustrate the system of engineering education as a *consistent system*.

## 2. THE FIVE KONDRATIEFF CYCLES

Let us look at Watt and his steam engine. It started the first of four long economic development cycles as observed by Nikolai D. KONDRATIEFF (1892 - 1930). These cycles are, therefore, called Kondratieff-Cycles or Kondratieff-Waves. The four different cycles are characterised by the invention of certain new *technologies* which subsequently cause and trigger a new *economic* boom.

These four cycles are:
- the steam engine and cotton processing;
- the train and steel production;
- electricity and chemical process;
- the car and petrol processing.

Since about 1980, a new cycle of economic boom has started: the *Fifth Kondratieff cycle*. It was triggered by the invention of the transistor, followed by the laser and leading to new concepts of *information* as a science: object-oriented programming, Petri-Nets, etc.. These developments demonstrate global impact and dynamics as no other development in human history (see fig. 1).

This Fifth Kondratieff cycle is dealing with fundamental societal, environmental, cultural and mental changes and challenges. Schumpeter, 1912, even calls such changes *revolutionary*. Leo A. NEFIODOW (Nefiodow, 1990) recognises the common roots of changes in all these different areas. He successfully analyses the interactions, trends and obstacles of these changes. Some of these observations are discussed in the following paragraph.

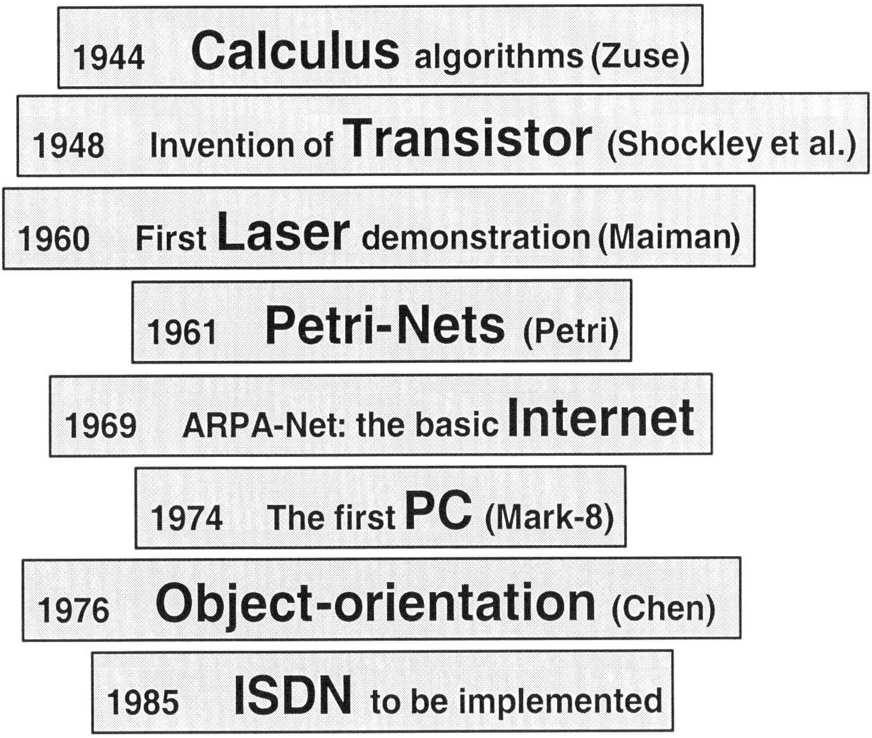

| 1944 | **Calculus** algorithms (Zuse) |
| 1948 | Invention of **Transistor** (Shockley et al.) |
| 1960 | First **Laser** demonstration (Maiman) |
| 1961 | **Petri-Nets** (Petri) |
| 1969 | ARPA-Net: the basic **Internet** |
| 1974 | The first **PC** (Mark-8) |
| 1976 | **Object-orientation** (Chen) |
| 1985 | **ISDN** to be implemented |

*Figure 1. Basic innovations in Information Sciences.*

## 3. IMPACTS OF THE FIFTH KONDRATIAFF-CYCLE

Around us, the main technologies are visible which are at the roots of all present changes. Even biotechnology is already well established with its global impact.

The engineer is in the centre of these *technological* changes. They can be characterised by certain terms, e.g. non-linear relations, networking, holistic structures, sensor-actor systems, modularised and re-usable software etc.. In the area of *economy*, the developments have caused mainly the transition from the seller-determined market to the buyer-controlled market. It has led to the following changes: e.g., global integration of enterprises and banks, international and regional outsourcing in hardware and software production, multi-dimensional strategies in control technologies, new dynamics in structuring multi-project management, needs analysis as the basis of production decisions, etc..

In *politics*, new quality awareness can be observed world-wide, also new environmental awareness and deepening safety awareness. The end of socialism has caused new business orientations of international enterprises and in parallel, it has increased the societal burdens of unemployment which need to be dealt with.

New patterns of work are emerging (e. g. tele-work). Furthermore new ethical challenges are becoming visible which need to be dealt with in a democratic way (Polke, 2000). Finally, new *educational* structures are being asked for. Hence some links of information and education are discussed in the following paragraph.

## 4. STRUCTURES OF INFORMATION

Through recent research, we have learned that humans are easily overburdened if information in writing or pictures etc., is offered in a linear, unstructured way. Information needs to be *structured* in order to be useful and usable. This view is particularly important in education. One model structure of information is offered by the approach of *object-orientation: cutting-up complex* information into smaller entities which are better comprehensible for the human mind. Furthermore, *similar* components of information can be *summarised;* in order to correspond to *view-orientation* approaches. Other information can be *modelled* to correspond to the approach of *process-orientation (see fig. 2).*

| Architectural-principle | Decompostion | Abstraction and Detailing | Transformation |
|---|---|---|---|
| Modelling-concept | Object-orientation | View-orientation | Flow-orientation |
| Examples for Modelling-tools | - E/R*-Model<br>- OMT** | - Level- Model | - SA***-Model<br>- Phase-Model of Production |

\* E/R = Entity Relationship
\*\* OMT = Object Modelling Technique
\*\*\* SA = structured analysis

*Figure 2. Information structuring.*

Beyond such *information structuring,* we need to consider *information presentation.* It means to *present* information in a way which more closely corresponds to the human capabilities of information processing. The *meaning* of information needs to be made accessible to the human operator: specifically if it concerns information which is hidden in the software of computers and computerised control systems (Ingendahl, 1998). The overt or latent hostility against technology within many groups of society has one of its gravest roots in neglecting this aspect of designing information technology. Hence the design of technological systems of any kind needs to take into account these fundamental aspects of human information accessing and information processing.

All human-computer interfaces need to follow the rule of *self-explanatory design*. They may integrate pictures, spoken language, noises and music. Furthermore they need to comprise the analysis of complex patterns (deductive approaches) as well as the pathways from the single detail to the whole system understanding (inductive approaches). The centre of concern is the *task to be performed* rather than the computer program.

"Those who know the hammer as their only tool, believe the world to only consist of nails".

## 5. INTERNATIONAL COMPARISON OF ENGINEERING EDUCATION

Engineering education at the German universities appears increasingly *detached* from the international patterns of education. Hence, the VDI and VDE have published several reports which emphasise the special features of German engineering education (VDE, 1997; VDI, 1998). They also contain clear-cut statements concerning the further developments of these educational structures in Germany.

The main feature of German engineering education is its one-way pattern. The university and the *Fachhochschule* (generally called *University of Applied Sciences)* are clearly separated from each other (see fig. 3).

In contrast, the Anglo-Saxon approach to engineering education allows any kind of *cross-over* between different courses of studies of the different universities and faculties. Therefore, VDI and VDE have suggested to make the system of Higher Education more open and flexible in horizontal direction (see fig. 4). This change process would need to begin with looking at *Secondary* Education in a different way. Thus the first step of change may be to start all Higher (or Tertiary) Education after 12 years of schooling (presently in Germany: 13 years as the standard Primary-Secondary Education). Within this Secondary Education, increasing emphasis should be on basic natural and engineering sciences according to the VDI present proposals. These proposals are based on about 30 years of commitment of the VDI to increase science and engineering teaching in *Secondary* Education. Subsequently, T*ertiary* Education of engineers should begin with a 1-year orientation course which leads to a new kind of university entrance examination. The course content would emphasise natural sciences and engineering. It would equalise differences between the university and the Secondary School teaching outcomes because these outcomes are fairly inhomogeneous across different types of schools.

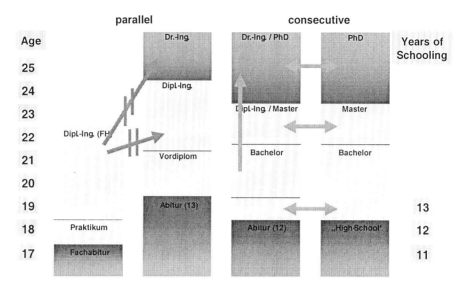

*Figure 3. International comparison of engineering education.*

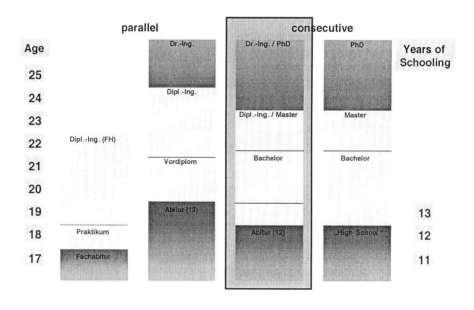

*Figure 4. The proposed comprehensive structure of engineering education.*

After this orientation year, university studies are to be structured more or less similarly to the Anglo-Saxon model of the *Bachelor-Masters-Doctorate* sequence. According to the discussions of the VDI/VDE, only such a consecutive structure of engineering studies can overcome the problems caused by the almost insurmountable barricades between the different courses of studies of the different types of universities in Germany. It will also make it easier for foreign students to master the German university education and its intricacies. Additionally it may contribute to solving some of the present problems of accreditation. In the next paragraph, the future structure of the university will be discussed as it is to be optimised in view of such fundamental changes of the university system. It will be a challenging task of Higher Education policy to develop these concepts further with the aim of implementing them through democratic strategies (see fig. 5)

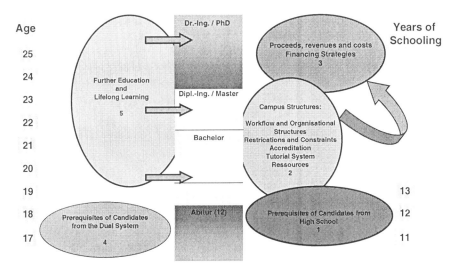

*Figure 5. Strategies of university policies.*

## 6. THE *CAMPUS* STRUCTURE OF THE UNIVERSITY

The *campus* structure of the university as described here, has been suggested by the VDI expert working group mentioned before; the description of this model follows the object-oriented approach because this approach supports analysing the university as a consistent system (see fig. 6).

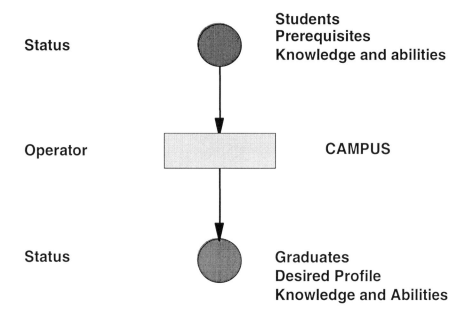

*Figure 6. The concept of transformation through university studies.*

Within this object-oriented approach, the first class to be considered is the class "state of education" which comprises the *students* passing through the system. Their main attributes are the "desired profile" and the "qualification profile". The attributes also comprise the entrance qualifications and the different final examinations and their required knowledge and abilities (Hillmer et al., 1976). All these aspects of the class "state of education" have been dealt with by the VDI, during past committee work. The most recent task concerns the profiles of the new class "campus".

The class "campus" comprises the *personnel*, the *teaching/learning methods* and the *economic resources*. The resulting tasks mean to work out the attributes of the class "campus".

With reference to *personnel,* the future qualification profiles need to be defined. Furthermore strategies need to be developed how to master the transition from present to future qualification profiles. One crucial issue is the requirement of the *"Second PhD"* (in German *"Habilitation"*) which is still the main entrance qualification requirement for about all professorial appointments in German universities. Another such issue is the whole process of *appointing professors.* It may be necessary to restructure this process according to today's standards of personnel management. Furthermore it relates to the departmental management itself which frequently shows a discrepancy or gulf between the management as such and the supervision of research. One solution was suggested already by Moessbauer (about 40 years ago): to introduce more non-tenured, non-permanent contracts into university management in order to allow for more flexibility in *task-related* appointments. It may include to make transitions easier between university, industry, enterprises or public administration.

With reference to *methods of teaching and learning,* the Working Group "Informatics" has put forward several proposals (GKI, 1997). They include scrutinising and evaluation in-depth today's methods of teaching and learning. Furthermore it means to evaluate the different teaching approaches as well as the roles of the teachers, the problems of content overload in teaching and the need to improve motivation of both teachers and students (VDI, 1999/2000).

The analysis of both workflow and organisational structures of education is an important task within the issue of teaching and learning methods. It may be helpful to compare the management of *educational* processes with the management of *enterprise* development. Furthermore it includes the questions of whether university is to put emphasis on educating *elites* rather than *average graduates.* The issue is to be dealt with of how and where to best *localise* the campus and its different – and largely independent – components and entities (see fig. 7). It includes the need for interactions of the university with the region around it. It also includes the *governance* of the university which today is mainly a State government task. It may need new structures of control mechanisms and university presidency overcoming today's well-established traditional structures.

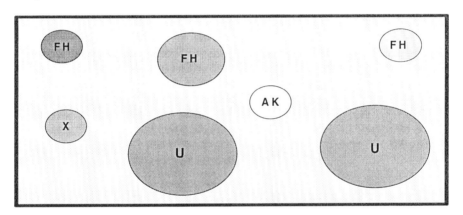

*Figure 7. The independent educational institutions within the framework of the virtual campus.*

With reference to *economic resources,* new financial strategies are needed to run universities in a similar way as enterprises. University *budgeting* today requires to make visible the real costs of infrastructure including personnel, materials, energy etc.. It also needs to show the proceeds, fees, subsidies and grants. The traditional budgeting approaches do not really fit into the 21$^{st}$ Century any more.

Beyond the issue of budgeting, the fundamental questions needs to be dealt with: how to *finance* the university of the future. There are the different tasks of the university which imply different approaches of financing: the *educational* task which may need to be re-considered in view of university fees (so far not yet common in Germany); the task of *fundamental research* to be funded by industry in co-operation with public funds; the projects of *applied research* which already today are mainly financed through industry contracts. *Venture capital* may be a fourth promising way of funding new applied research projects

## 7. FURTHER EDUCATION AND LIFELONG LEARNING

As mentioned before, society is challenged fundamentally through the global changes and the technological developments, Lifelong Learning is becoming a basic requirement for everybody. Society needs to activate gigantic efforts and investments in order to cope with this challenge. Most traditional educational institutions are not prepared for it. The university may need to support this process. It may need to quickly respond to such new needs following thorough analysis of these needs, the persons in need, and the contents needed. Such response may lead to new roles of the university within society. The VDI has put forward the following issues shaping future Further Education programmes and Lifelong Learning approaches (VDI, 2000):
- the technological developments (specifically information and computer sciences) which are shaping the transition from industrial society towards information and knowledge society;
- the global competition replacing the regional competition of the past;
- the markets which are characterised by the transition from seller-orientation to buyer/user-control;
- the degree of automation which is presently re-defined by the transition from industry/production orientation towards service orientation.

The participants of Further Education programmes may be fully employed aiming at promotion or at keeping their jobs; they may be presently unemployed requiring re-education in order to become again employable. This latter group needs particular emphasis in university-based courses. The gains of such programmes must be long-term advantages for these participants rather than fast and short-lived learning outcomes. Examples of course contents are shown in fig. 8. The teaching of such long-term contents poses additional challenges for all teachers within these programmes.

## 8. CONCLUSIONS

This report has dealt with a wide area of issues and concerns. The main issue has been to describe the engineer of today and tomorrow in the transition of industrial society towards information society. Technology and economy within their global framework are defining the most important scenarios of the future of engineering professions. The university is expected to respond to these scenarios. The response, however, needs to be based on considering university policy as a *consistent system*. Thus the challenge for university policy is to search for its own path into the future by re-defining engineering both as a science and as practice and application. Furthermore, university policy needs to evaluate every step it goes, in terms of present and desired states of its workflow and organisational structures and its methods of teaching and learning.

> - Information technology
> - life-cycle qualification for all engineering tasks
> - requirements orientation and marketing
> - new materials and processes
> - new approaches of organising human work
> - project orientation, team work, tele-work, international networking
> - foreign languages teamwork - entrepreneurial competencies

*Figure 8. Proposed context of Further Education for engineers.*

## 9. ACKNOWLEDGEMENT

The author would like to thank Dr. Dietrich Brandt for his support in preparing the English version of this paper.

## 10. REFERENCES

Polke, M. (1988): Information als kritische Ressource. Henzler, H.A. (Hrsg.), Handbuch der strategischen Führung, Wiesbaden: Gabler-Verlag, S. 353-378.
GKI (1995): Gesprächskreis Informatik: Informationskultur für die Informationsgesellschaft. Druckschrift des Gesprächskreis Informatik, Frankfurt.
Forum 2000 (1998): Bildung und Medienkompetenz im Informationszeitalter. Druckschrift der AG4 des Forum 2000.
GKI (1997): Gesprächskreis Informatik: Informationskultur für die Informationsgesellschaft durch Bildungsinitiative Neue Medien. Druckschrift des Gesprächskreis Informatik, Frankfurt.
Nefiodow, L. A. (1990): Der fünfte Kontratieff.: Strategien zum Strukturwandel in Wirtschaft und Gesellschaft. Wiesbaden. Gabler Verlag.
Polke, M. (2000): Informationskultur für die Informationsgesellschaft: Brauchen wir eine neue Ethik? Steinheimer Gespräche des Fonds der Chemie für den Hochschullehrernachwuchs, Mai 2000.
Ingendahl, N. (1998): Konzept zur Analyse der Aufgaben und Auswahl der Präsentation für die Mensch-Prozeß-Kommmunikation. Dissertation, RWTH Aachen, Lehrstuhl für Prozeßleittechnik, Februar 1998, Mainz Wissenschaftsverlag, Aachen.
Färber, G.; Polke, M.; Steusloff, H. (1985): Mensch-Prozeß-Kommunikation, Vortrag auf dem Jahrestreffen 1984 der Verfahrensingenieure, München, 19.21.09.1984, Chem.-Ing.-Techn. 57, 4, S. 307-317.
VDI (1998): VDI Thesen zur Weiterentwicklung der Ingenieurausbildung in Deutschland, Düsseldorf.
VDE (1997): VDE-Tätigkeitsbericht 1996. VDE-Öffentlichkeitsarbeit, Frankfurt.

VDE (2000): VDE-Studie: Ingenieure der Elektro- und Informationstechnik, Frankfurt.
Hillmer, H, Peters, R.W., Polke, M. (1976): Studium, Beruf und Qualifikation der Ingenieure, VDI-Verlag, Düsseldorf.
VDI (1999/2000): Diverse Verlautbarungen zur Akkreditierung, VDI Hauptgruppe "Der Ingenieur in Beruf und Gesellschaft", Bereich Aus- und Weiterbildung, 1998 ff., Düsseldorf.
VDI (2000): Weiterbildung, Versuch einer Bedarfsanalyse, VDI Hauptgruppe "Der Ingenieur in Beruf und Gesellschaft", Bereich Aus- und Weiterbildung, Ausschuß Weiterbildung, 1999/2000, Düsseldorf.

*Martin Polke*
*Aachen University of Technology*
*Aachen, Germany*

Y. J.M. BRÉCHET

# INTERDISCIPLINARY TRAINING FOR ENGINEERS - A CHALLENGE BETWEEN SUPERFICIALITY AND OVERSPECIALIZATION

**Abstract.** The need for cross disciplinary training for future engineers is discussed. The inadequacy of the classical classification of scientific disciplines and of the standard ways of teaching is stressed. Examples of "problem setting/problem solving" approaches from project learning are proposed. The necessity of not building barriers rather than trying to remove them is emphasised. The advantages for interdisciplinary training of including non scientific disciplines in the curriculum are discussed.

## 1. INTRODUCTION

Interdisciplinarity, and cross disciplinarity, are key issues in the training of engineers. It has always been the case, and what seems nowadays a goal to reach was an obvious aspect of the competence of Renaissance engineers. Are we again addressing one of these trivialities that pedagogical discussions seem so generate so naturally ? It may be the case, but nevertheless the evolution of the requirements on the training of the engineers, the variety of the disciplines involved, as well as their autonomous developments, the emergence of engineering fields relying on collaborations between disciplines, urge us to address this question, not simply in terms of wishful thinking, but in terms of practical and economically viable solutions for our universities.

## 2. THE DANGERS FACING THE TRAINING OF FUTURE ENGINEERS AND THE NEED FOR INTERDISCIPLINARY TRAINING

Training future engineers is nowadays facing two dangers : over specialisation and superficiality : these two dangers are real ones, and some of our formations have certainly experienced some of them...

In the first case (over specialisation), we would be training engineers knowing everything on almost nothing : they might be excellent specialists of a very narrow technical domain, very efficient, but deprived of the intellectual tools enabling them to face the rapid evolution of technologies, or to borrow from other specialists possible solutions for their own problems.

In the second case (superficiality), we would be training talkative well-to-do nothing, a specie which might have some brilliant careers in some companies specialised in hiding technical incompetence behind prestigious managerial abilities. It might be a trend in western industrial societies to forget where lies the root of their wealth - technical abilities to create, make and sell industrial products - but it is a duty for the universities and their professors to be sure that they are training engineers able to create or produce something, and not only to talk about production and creativity.

Interdisciplinary training is not a fashion : it is at the very core of the engineering discipline. Since the engineers are supposed to <u>create</u> industrial products, to <u>realise</u> them, and to <u>sell</u> them, their training will require a vast range of competence. Depending on the emphasis in their career (and often depending on the period of their activities) the engineers will have to rely on a solid scientific and technical training, to practise managerial tasks in the production, and to face the economical and marketing challenges of a competitive industrial world. All along his career, an engineer will have therefore to acquire and integrate highly technical and more "experienced based" competence. It seems obvious that the knowledge required to make a good engineer cannot be acquired as a simple sum of competence, and that only during their initial training. Even assuming that universities would be able to train engineers as universal specialists of science, technology, production engineering, economists, it is totally unrealistic to expect them to acquire at a usable level such a wide range of skills, and the superficiality which would result from the illusion of the opposite can please only the fields of activities looking for « varnished engineers ». Requiring that properly trained engineers should be able to learn these competence during their careers, and should have the intellectual tools to do so, is a real and realistic challenge, and that is the true aim of interdisciplinary training. The program proposed in the present paper is « training excellent specialists able to collaborate with other excellent specialists ».

We will focus our attention on technical training. It doesn't mean that we underestimate the importance of behavioural qualities compared to technical knowledge. It means that these abilities (such as working in group, communicating, practising foreign language, being exposed to other cultures), can be learned from practise rather than from formal lectures. Nothing seems to us more ridiculous than a « course on creativity », or « a credit on spontaneity » : these qualities are inherent to the engineering practise, and we claim that it is through practise that they should be acquired.

## 3. THE WEIGHT OF TRADITION IN CLASSICAL ENGINEERING TRAINING

Engineering practice has always been forced to evolve, to improve, thank to the economical sanction : more and more, efficiency is a condition of survival. The competence of engineers have enormously evolved , but the qualities of a good engineer remain the same as in the time of Effie, Brunel, Ford or Edison : a good grasp on the technical aspects, a practically oriented mind, a balanced mixture or analytical thinking and empirical knowledge.

By contrast, universities are probably among the most conservative of human institutions. They have been evolving a lot according to the <u>content</u> of the knowledge they produce, store and distribute, but the often still rely on two obsolete hierarchies which are the basic of the curriculum content and of the pedagogical practices.

One of these hierarchies is the classification of sciences by Auguste Comte, from the more basic ones to the most applied ones, from the simple concepts to the complex objects. In this classification, physics is seen as a branch of mathematics, chemistry as an application of physics, biology as an offspring of chemistry. This caricature of scientific activities doesn't survive to any serious practise of any of these sciences, but nevertheless the curriculum, and especially the undergraduate curriculum, still relies strongly on this misconception of sciences. It is clear that a strong training in basic sciences is needed for the future engineers, but the very first condition to instil in his behaviour the intellectual ability for cross disciplinarity is to get rid of this hierarchical view of fundamental sciences.

The second hierarchy is more insidious, since it undermines the practices and not simply the curriculum. The hierarchy of pedagogical practices lists in order of decreasing importance the formal lecture, the tutorials, the practicals and the application projects. This hierarchy of course flatters the natural tendency of professors to show off on the stage. It is usually justified on economical grounds, the amphitheatre lecture being able to train a larger number of students. The efficiency of formal lectures compared to tutorials is however very low, and the difference of cost should be estimated, not for the same amount of knowledge provided to the student, but for the same amount of knowledge acquired by the student. A good formal lecture aims at giving to the student the general ideas, the will to learn more, the interest to deepen his understanding, it has to motivate the future engineer, but the real knowledge, the practise of the concepts requires the tutorials, in much smaller groups. It is reasonable to admit that depending on the qualities and experience of a teacher, he might be more efficient in one form or another, but there is no reason to undervalue the tutorials compared to the lectures. The underestimation of the value of practicals and project work is even worse that the traditional prestige associated to the lecture compared to the tutorials. It gives the student the disastrous misconception that real life is a mere application of what he has been taught in class. It is crucial to train the future engineer to interdisciplinarity, to help him to discover that knowledge is built from problems, and not that problems are exercise to apply knowledge.

The two basic hierarchies which structure the classical training of engineers are instilling in their mind two wrong ideas : that the various basic disciplines are related in a hierarchical way, and that practical problems are application of a well defined body of knowledge. Interdisciplinarity requires on the opposite to be able to understand the specificity of the basic disciplines, and the need to involve several of them in the analysis and hopefully the solution of a real engineering problem.

One might argue against this rather frontal criticism of classical structures of engineering education that very good engineers have gone through this training, without too much damage to their brains. It only proves that an engineer is not simply a puppet produced by his professors, and that engineering practice corrects the drawbacks of engineering training. It also proves that the partitioning of knowledge and industry which was typical of the 19$^{th}$ century and the first 50 years of the 20$^{th}$ could live with this double hierarchy. It doesn't prove that the acceleration of technical development, the new fields at the margin between classical disciplines ( telecommunication and computer science, mechanics and materials, process engineering and chemistry, biotechnology...) can be developed in universities in the same framework. Interdisciplinary thinking requires effort not only on the curriculum, but also on the underlying intellectual ideology behind our pedagogical practices.

Recent years have seen an evolution in teaching practice : the academic fossils of « exercises of application of the course » has left, and has been hopefully replaced by a « problem solving approach » allowing a more synthetic view on the engineering competence : this was certainly an important first step toward interdisciplinary training. We have now to evolve toward a « problem setting approach » which would complement the « problem solving approach », teaching the students the complexity of an engineering problem, and the harmonious mixture of analysis, induction and experience, the need for collaboration between different expertise, which will be their real life as engineers in industry.

## 4. AN EXAMPLE OF INTERDISCIPLINARITY TEACHING DESIGN

An example of new requirements in training engineers is provided by the teaching of design. Lets focus for a moment on structural design: this encompasses industrial fields such as automotive, aeronautics, machine industry.

Product design starts from the identification of a requirement of the market, and the investigation of the possible concept to fill this need. Inventing the ball pencil, or the optical fibre requires both an understanding of the market needs, but also an imaginative mind able to invent new concepts. After the specification of the concept, the design of the object has to go through conceptual design, embodiment design and detailed design. The skills needed at this level are mainly scientific skill, such as mechanics, materials science. In order to be able to guide the choice process in this design procedure, the engineer will need to master new computer tools, but above all will need to have a comparative mind. Selecting the materials and the process able to realise the object requires competence from technology as well as from science. But the story doesn't stop there: the object has to go into production. Minimum skills in economics are needed to have an appropriate business strategy which allows to evaluate the technical and economical viability of the product. In the case of mass production, very often industrial design, aesthetics, and more recently environmental concern and sustainable development will be important aspects of the design procedure: a product is nowadays designed for production, sale, use and disposal, and the design engineers needs to have all these aspects in mind if he wishes to design a successful product.

Does it mean that we need to overburden our student with lectures on materials, mechanics, economics, marketing, eco-design, and aesthetics? Does it mean that we have to fill his brain with engineering sciences, production engineering, human sciences and fine arts? More efficient, we think, is to give him the basics in engineering sciences, the mechanics, the materials science required for a technical design to be feasible. Then, this theoretical knowledge has to be digested by practice: instead of exercises, the student will have to do a project, given by an industry, for which he will have to propose a solution in terms of materials and processes, the "in house feasibility" will be investigated with the industry which gave the project. The economical requirements will naturally come via the interaction with the engineer who is tutoring the project. The students will work in teams, formed of students of different competence, one of them might be a designer and not an engineer, one of them might be a student in marketing. The engineer which has gone through this training by project will not have a fake competence in economics or in marketing, he will be above all a technical engineer, but he will be able to interact with other people in the company who will hire him. He will have learned not only to solve a problem, but to set a problem, to start from a vague requirement and to make it more precise, technically tractable, by interacting with the industry who asked for the project. We have been practising this method in INPG for training in design, students coming from departments of mechanics, physics and physical chemistry working together. Some of the projects were performed in collaboration with furniture designers, others with marketing students. The industry providing the projects, and following the progress of the students were materials producers or materials users, were large multinational companies as well as small and medium size companies. The variety of projects treated with success, the enthusiasm of students to learn from real life, to think of their lectures not as source books for predefined solutions, but as toolboxes for them to create their own solutions, is really encouraging for the possibility of having an efficient cross disciplinary training.

We claim that this experience in mechanical design can be done in many other fields of engineering sciences. However, in spite of the difficulty of collaboration between, say, mechanics, materials science and process engineering, the cross disciplinary training in this respect is possible : it requires simply professors of good will, experts in their own field and ready to collaborate with other experts. The "project setting approach" is a strategy which is possible only with a faculty ready to be as cross disciplinary as the student they want to train.

## 5. AN OPTIMISATION PROBLEM UNDER CONSTRAINTS: WHAT ARE THE MOST APPROPRIATE PEDAGOGICAL METHODS ?

The case of teaching design is not specific. A number of fields need cross disciplinary training. To name only a few of them, electronics and computer science, process engineering and chemistry, environmental engineering, biology and electronics, biology and computer science, bio-mechanics, will correspond in a very near future to huge demands from industry. Is it thinkable to create in our universities as many structured curricula as would be required by all the possible combination between various disciplines? This seems impossible for practical and economical reasons. The modern students still have only 24hours per day, and not all of them are devoted to work!

The basic disciplines have to be taught at the best possible level. The structure Lecture/Tutorials/Practicals is a reasonable and economical way of providing the basic knowledge.

Some "combined fields" such as design can be taught in a structured manner, involving lectures and tutorials, but mainly project work between students of different origins. This type of cross disciplinary training is relatively easy to implement, provided both the faculty and industry are ready to take the risk, on one side to look for real problems, on the other side to invest time into training students from example. At the level of the faculty, such fields are intellectually reasonably close so that any professor in materials should have some interest in mechanics, any professor in structural mechanics should have a basic curiosity in materials, any professor training engineer should have once felt the concern about economical aspects of engineering. It is therefore relatively easy for a faculty which is not fully ossified in narrow-mindedness to set up a curriculum on mechanical design.

Setting up a curriculum in bio-computer science for instance is considerably more difficult since the basic disciplines ( rather young and evolving rapidly) have very often not even met at the level of faculty. The solution in that case is not so much, we think, in setting up a defined curriculum, but in allowing this possibility for a limited number of motivated students who, for reasons from their personal experience, got interested in these unexpected links between very different disciplines. A number of emerging and promising fields will require new expertise: a way for the universities to build this expertise is to develop a system of close interaction, for very motivated students, with some professors ready to help them to build these competence by a training "a la carte". This might sound a very elitist approach, and indeed, it is. But it is a pragmatic one too: in some fields, cross disciplinary approaches are very new, and the only way for a professor to get some experience is to make the path together with the students he has to train.

## 6. STRATEGICAL APPROACH: AVOID BUILDING FRONTIERS RATHER THAN HAVING TO REMOVE THEM

Up to now, we have examined in this paper the obstacles to cross disciplinary training in terms of pedagogical practice and discipline classification. We have outlined some possible ways to overcome these obstacles, insisting on the necessity of a "problem setting approach" in relation with projects. A more radical approach to bypass barriers between disciplines is "avoid building frontiers rather than having to remove them".

Frontiers between disciplines : the plague of cross disciplinary training is to fight the need of "labelling" the problems. Is it physics? Is it chemistry? Is it mechanics? Is it biology? Underlying this need to label the problem, there is this unsaid belief that real problems have solutions to be found in some lecture notes. A simple and efficient way to protect the students from this ideology is to have them involved, very early in their training, in practical problems, preferably unsolved, for which they would have to design an experiment, to observe a problem, to find a possible explanation and to test it. Number of problems can initiate with the students the will to look around with a scientific and engineering insight: how does a spaghetti break? what makes the ball bearing of a roller blade fail? What controls the size of the hydraulic jump in a sink? What is the fine structure of a straw? How is an eggshell structured such as the bird can break it but only from inside? What makes the oyster shell so tough? What controls the life time of a light bulb? How can one recognise a fake coin? What is the collapse velocity of a domino stack? What makes a water flow meander of a window? How do the ripples form in the dunes? What are the growth mechanisms of bumps on a skiing slope? Why lemon juice stabilises egg foams? It is very easy to find many everyday life questions on which a young imaginative mind can have a try, without being obliged to label the problem, to categorise it in predefined field of expertise. The student need to have a structured teaching of the various scientific fields in engineering sciences, and that is perfectly acceptable. But they also need to be taught that, beside scientific knowledge, scientific method is as important for their future life ( as engineers as well as scientists). Close observation of nature, setting up experiments, finding qualitative explanations, writing a few equations on the back of an envelope to test them should become a natural reflex for them. It is an efficient to have them thinking in terms of problems , and methods to attack them, rather than in terms of disciplinary fields that would need to be stacked together in a somewhat artificial manner. Moreover, it is important that engineering students have very early in their studies the feeling that they can be actors in the understanding of nature, and not simply receptors of a pre-existing knowledge. Once again, a project approach, working in small groups, in collaboration with one of their professors, on problems for which the answer is unknown, seems to us the most efficient way of developing this natural curiosity that any intelligent student has.

Frontiers between activities: as surprising as it may seem, there is a natural tendency among faculty to practise intellectual cloning on the students! Instead of training engineers, some of us tend to train people who would train engineers (possibly)...A unsaid hierarchy would be to train the very best students to become academics, and the less good ones to become engineers in industry. The ones who are smart enough to play with concepts should remain among the great priests of knowledge! As can be expected , the reaction is , for the ones who have gone to industry, to despise the conceptual knowledge which was presented to them not as a way of understanding and mastering nature, but as a way of sorting the different classes of students! Often, the deficit in communication between industry and university reflects this unsaid superiority/inferiority complex. This is always an unpleasant situation in traditional disciplines, it will become, if we don't pay attention to it, a dramatic one for emerging fields in which the science and technology content, as well as the rate of its evolution, is so enormous. The divorce between industry and university in traditional disciplines is a problem for both. In new fields such as advanced computer sciences, or genetic engineering, the divorce might be a disaster for the universities: it is already quite obvious that in these very competitive fields, the frontiers of sciences are not reached in universities but in industries. It is absolutely necessary to have the best people from industry coming in our universities for conferences, for tutoring our students, as well as for our best faculty members to collaborate with industry, to act as consultants, so that the choice of the students to go to industry or to remain in the academics will be a real choice from their abilities and their taste , and not a simple reproduction of the existing academic structure.

Involving students in "thinking projects" ( more general than research projects) to teach them scientific method rather than locking them in a disciplinary approach, involving industry in the training process and faculty in engineering thinking, to have a better communication between academics and industry, these are some practical ways of NOT building barriers rather than trying afterward to remove them.

## 7. SCIENCE AND TECHNOLOGY AS CULTURE

A key issue in the training of engineers , and more specifically in the challenge of giving them the intellectual tools enabling them to build continuously interdisciplinary competence , it to forward to them an image of science and engineering as an integral part of the culture of their civilisation. They must not feel they are just here to provide society with worldly goods, comfort, energy, power... They should be aware that they live in a world which is shaped by engineers, and that not knowing the basic ideas of quantum mechanics or the way a light bulb is made is a sign of illiteracy , as well as being unaware of Shakespeare or having never heard of Bach. Student engineers should be given this feeling that, although they cannot be competent in everything, they have to be curious of many things. They should be given an image of science and technology as a living activity, as a way of understanding and reacting to the surrounding world. They should be taught to look around and wonder « How is it made ? How does it work ? Why is it so ? ».

This inquisitiveness of mind can be given to the students in the way we teach them engineering sciences, it can be proposed as a way of entering into the various technical projects described above. This approach will certainly enhance their ability to « lateral thinking ». This requires from the professor themselves that they complement their technical knowledge of their own discipline with a minimum awareness of the other branches of science and technology, of their methods, of their questions. There is certainly a parallel action to be taken, how can we train properly ourselves to be more than overspecialised experts ?

But another way of developing this understanding of science and technology as cultural activities is to provide the students, in their curriculum, with contacts with literature, history, philosophy, social sciences. Very often , we (science professors) have a condescending utilitarian view on the « non scientific training » of engineering students. The result is the implementation in engineering courses of « communication skills » which, at the very best give the student some rhetorical abilities, and at the worst give them the unpleasant and unjustified feeling of emptiness of social sciences. This utilitarian view of humanities is the worst possible way of implementing non scientific topic in engineering curriculum. It has led to the situation where it is felt that only economics ( and microeconomics since it looks more « scientific » with its thin layer of mathematics ) and communication are worth teaching to engineering students. Humanities are viewed then as « technical knowledge » with words, or as a « supplementary soul » which would enable the future plant director to brilliantly entertain the audience in an evening reception at the city hall...This is , to my opinion, a total misunderstanding of the role of humanities in the training of engineers . Humanities are absolutely necessary to give them the conviction that their activity in society is an important part of the civilisation process ( history of science and technology, epistemology can play this role ). They can teach them that communication between cultures is an important aspect of their activity (comparative literature may be an excellent way of learning that). They can learn from them that being able to follow a reasoning in a totally different field, when someone makes it understandable, is a prerequisite of intelligence and creativity.

We have to train our students as independent minds, as citizens, as men of culture. That has to be done in close collaboration with our colleagues in humanities departments, those at least who are not deeply convinced that engineers and scientists have nothing to learn from them or nothing to teach them.

Our generation has seen the divorce between the « two cultures », it is our duty to help our students to heal this would in the body of our civilisation, that is part of their training as efficient engineers. We have , as scientist , to show them that there is one generic scientific methods, and many sciences, but that being a scientist and an engineers is also being a man of curiosity : they have to be convinced that narrowness is not a condition of efficiency, and rather a sure method toward mediocrity. We have, as citizens and human beings, to convince them that their training as engineers is part of their training as complete human beings, and that they don't need to be schizophrenic to be interested in sciences and in humanities. Humanities are not the polish one puts on the shoes to make them shine, its also the core of the training of engineers, the ground on which their future ability to develop as balanced citizen will grow.

## 8. CONCLUSIONS

The obstacles to cross disciplinary training outlined in the present contribution are identified from the experience of the author, who, in spite of numerous experiences abroad ( mainly in Anglo-Saxon countries) has been mainly teaching in France. However both the inertia opposing to interdisciplinary training, and the possible solutions appear to be not so specific of a given country. The author is a professor in materials science, which by nature is cross disciplinary. Most of the examples given in the text will come from this field. However, it seems to us that the various solutions proposed to face the need of cross disciplinary training are also not limited to materials science. From our personal experience in training engineers in a specific educational system, we have tried to outline some ideas which, we hope, are more generic than the examples which motivated them. Cross disciplinary training for engineering student is a key issue for the present and the future. It is meaningful only when associated with an efficient specialised training, and its aim should be to enable our students to be good experts able to collaborate with other experts. This requires a profound evolution in our pedagogical practice, it demands to abandon the obsolete positivist classification of sciences, it necessitates a diversification from the classical scheme "course / tutorials / practicals" to more project based learning with an active tutoring. It implies a closer relation between academics and industry, and a better understanding of the possibilities offered by social sciences to help us make the students and ourselves realise that science and engineering is part of the culture of our civilisation, and not only a way of earning a living. This evolution in our teaching practice is a necessity if we want to be able to train the engineers of the future, all those who have started in this direction and have experienced the enthusiasm of students being made responsible for their learning through project work know that this necessity is also a real pleasure well worth the time invested in it.

## 9. ACKNOWLEDGEMENTS

It is a pleasure for the author to thank many colleagues for stimulating discussions over the years on the question of training engineers , among which M.Ashby, D.Embury, R.Fougeres, P.Guyot , J.C.Poignet, and G.Purdy. The recent experiences in novel ways of training engineers in INPG have greatly benefited from active collaboration with younger colleagues , A.Deschamps, D.Imbault, Z.Neda, L.Salvo, M.Veron. Over the years a number of industries have helped us in providing projects and guidance. Finally, it is always a pleasure for someone who trains students to recognise how much his own way of teaching owes to his own professors, and it is a pleasure to acknowledge the initial and continued influence of Y.Quere. Needless to say that the non diplomatic statements of this paper are the author's responsibility only!

*Y. J.M. Bréchet*
*Institut National Polytechnique de Grenoble*
*L.T.P.C.M. / E.N.S.E.E.G.*
*Saint Martin d'Hères, France*

# PART III.

# ENGINEERING EDUCATION IN EMERGING ECONOMIES

R. N. BHARGAVA

# PRESENT ENGINEERING EDUCATION IN INDIA
## - AN EMERGING ECONOMY -
## AND A GLIMPSE OF THE SCENARIO
## IN THE 21$^{ST}$ CENTURY

**Abstract.** After gaining independence from Britain in 1947, Prime Minister Nehru conceived the vision of present day engineer and engineering education for the ultimate economic development of the country. He immediately asked the planners and educationists to plan and establish engineering institutions of various levels all over the country with the help of willing developed economies like USA, Germany, U.K., and Sweden.
The first three decades saw coming up of engineering polytechnics, regional colleges, national institutes and universities in India, that led Indians emerge at present as a leading group of people from a country, earlier an underdeveloped one but today one of the top in the world in the field of computers, software, internet etc., expert engineers and biggest entrepreneurs operating in the leading economies.
Hitherto emphasis had been on producing theoretically taught engineers with focus on theory and research aspects, sidelining more emphasis on practical training and educating them as a global engineer. This concept is, now for future being changed. More and more new institutions are being set up by Indian billionaires living abroad, which are going to be citadels of Global Engineering Education in the 21$^{st}$ Century.
The talk covers the history of engineering institutions and education since independence up to the present day and visualizes the scenario in the millennium.

After achieving independence in 1947, the Indian leaders in the government and the planners immediately realized the importance of developing engineering education in the country to ultimately build its industry, roads, dams, communication system, power and drinking water facilities and other infrastructure in general. This is the foundation for improvement of overall quality of life of people and to raise the living standard of nation. The engineers are the backbone and form the core of a nation to enable it become a leading country in the world.

Prior to independence, the country had handful of engineering colleges. On an average, each province had one or two, where only graduate level education leading to the award of B. E. degrees was imparted in the fields of civil, electrical and mechanical engineering. The first three years until 1950 were the years of planning and thereafter, the country entered into an era of establishment of national, state or regional, and divisional level engineering institutions mainly for graduate courses. Slowly over a decade, transformation for postgraduate engineering education set in.

The planners and the government also realized the necessity of practically focused and trained, lower level in theory, diploma holders in the above mentioned three fields to work under the graduate engineers. Therefore, one saw simultaneous setting-up and growth of so-called polytechnics. This sector also has been an important part of engineering education in India. The first four decades until 1990, have been thus a period of growth of institutions in above categories for imparting engineering education on the pattern of British or American systems. During this period more courses and disciplines were added in the fields of chemical, metallurgy, electronics, telecommunications, foundry, paper, sugar, textile and oil, in addition to aeronautical, mining, agricultural engineering, to name main branches of teaching. Mostly the country developed institutions for graduate level of education.

Middle of eighties saw the beginning of introduction of modern management techniques for manning the industry and engineering business enterprises. The promoters and managements of these institutions realizing the need and demand of this knowledge by the engineers of future in running various projects in different positions in industry and development projects, started adding faculties of business management on the lines of US in their institutions. These were in addition to sole-purpose business management schools.

From 1990 onwards, with the development of computer technology in USA, in India we started with the introduction of computer technology graduate level courses in Indian institutes, regional colleges and graduate level courses in others, set up by industrial houses. Thus during the last few years, the world has seen large number of computer hardware and software engineers from India. Simultaneously due to such engineers, who had migrated to US, who created the present Silicon Valley and in no time there came a period of information technology engineers from India contributing globally.

India reportedly has 662 rural and urban engineering colleges, and 1172 polytechnics having combined student strength of 357000. There are about 35000 engineering teachers, graduate and post graduates. There are 6 Indian Institutes of Technology and one IISc., Bangalore. There are one each national institutes of sugar, textile, petroleum, paper, oil, mining and score of institutions in agriculture engineering fields. Over 400000 computer, software and information technology engineers working in the world are Indians.

At this juncture, I consider it my duty to mention that in the task of developing engineering education and establishment of the institutions in India, many countries have provided much needed help and assistance. The positive roles played by USA, Germany, UK, Sweden and others need special mention. The signing of protocol by Prime Minister Nehru and Chancellor Adenauer in 1956 and subsequent opening of the DAAD office in New Delhi, to facilitate and organize exchange of academics in the field of engineering education was a milestone in the direction of developing highly trained engineers with global exposure, that led to building up of present citadels of engineering education. The Madras Agreement with Germany for IIT Madras; the help extended by US universities to IITs at Kanpur and Bombay; cooperation with UK for IIT Kharagpur are the fruits of their help and I thank on behalf of my country.

The institutions in the country are thus set-up by many different agencies viz. central government, respective state governments, several industrial houses, individual businessmen, religious temples, churches, Muslim wakfs and Sikh gurudwaras and are being run and managed as societies and trusts. The last two years saw several engineering institutions being promoted and established by IT billionaires and millionaires of US Silicon Valley. There are also ongoing educational projects of IT companies of India such as NIIT, APTECH, TATA INFOTECH, INFOSYS, WIPRO and SATYAM. These recently promoted engineering institutions are going to contribute to the scenario of $21^{st}$ century. Thus the last half century of post independence made India develop a large base of engineering institutions for almost all disciplines. There was some interaction between the teachers from other countries through score of foreign funding agencies but there has been little activity for undergraduate level students from western countries coming for their studies to India except from SAARC countries. To develop the needed global engineers, more efforts are required by all concerned so that this sector's activity is increased. More and more number of students should be sponsored by government agencies, trusts, societies and industrial & business houses. This would lead to better global cultural relations and understanding between the people in the world, particularly among the engineers of the future.

Now, I would like to conceive the likely scenario in the $21^{st}$ century in my country. On this issue recently Mr. Bill Gates gave his view in India that "we are living in very exciting times. The pace of change is dizzying and the impact this progress is having on our present and on our future is difficult to comprehend in its entirety. This is the age of powerful micro-processors, sophisticated software and new hardware technology. The universal connectivity will bring information and services we need and make them available to us where ever, when ever and on what ever devise we choose." This convergence of all the information will be the core for the Global Engineer of $21^{st}$ century. Gates wants the governments to use the latest technologies available to revolutionize their education systems, particularly for engineers.

The Ministry of Information Technology is seized of the situation due to phenomenal growth of the IT sector worldwide leading to exodus of country's engineers to US, UK, Germany, Japan, France and elsewhere. A task force constituted by the Ministry has recommended an investment of Rs 28 billions to improve the present infrastructure, improving computing and networking facilities, adding intake of students, training and retraining of engineering teachers as well. Madras based ANNA UNIVERSITY is going to be the first Indian Engineering Collage to impart post-graduate courses on the web. It has tied up with an Infotech company as a Joint Venture. It has plans to develop course contents with the help of other Universities and to operate in a manner that students from other states could also benefit from its online education.

The Minister of Technical Education of Andhra Pradesh State on $28^{th}$ September announced that his Government had decided to introduce electronic classes in all of its engineering colleges. It has planned to provide video conferencing and connectivity to World Wide Web, and querying and replying on E-mails. Thus in future the best resources and educational content globally could be developed and shared. In this context, the Chairman and Director of IIT Bombay have gone to USA to raise 100 million dollars from its old alumni's. Likes of Rekhi & Gupta of Silicon Valley have set up modern institutions by investing millions of dollars, for latest engineering education in different fields. The $21^{st}$ century will see vast developments in the field of genetics, neutracheuticals and biotechnology, requiring specialized, knowledgeable engineers in these disciplines. We already have National Institute of Molecular Biology, Central Drug Research Institute, National Botanical Research Institute to name a few in this direction, which are conducting valuable research with their applications requiring global engineers. Therefore, the syllabus and program for a global engineer will include focus on: The commercialisation of internet based content keeping in view the changing demands of the industry; adaptation to demands of internationalisation; developing personal skills to be a successful global engineer and entrepreneur and lastly on developing cultural relations with each other from this stage of life as a student.

During the $21^{st}$ century, an emerging economy like India would find many undergraduate and postgraduate engineering students from western and Far East including south-eastern countries. The cost of living and education is cheaper in India than in these countries. The hitherto process will be reversed, where our students and engineers used to study abroad. With medium of education being English, India would be an ideal choice for global engineering education. It is this platform, which should act as a torchbearer for promotion of such global order. This forum should focus attention to bringing multi-party cooperation of Government funding agencies, technical universities and multinational industrial houses to achieve this objective for $21^{st}$ century by sponsoring more and more number of students desirous of becoming a true GLOBAL ENGINEER. The Indian scenario should be thus clear to visualize.

*R. N. Bhargava*
*Germania Associates*
*Saharanpur, India*

S. M. CHEHADE

# LEBANON AS AN ENGINEERING EDUCATIONAL CENTER IN THE MIDDLE EAST

**Abstract.** Lebanon is a small country with a total area of 10,452 km$^2$. The population of about 3 millions is unevenly distributed, being concentrated mainly within the capital Beirut. The native language of Lebanon is Arabic, however English and French are the languages principally used in education especially in teaching Math and Sciences.

Lebanon has one of the best education systems of the Middle East, where, besides a modern pre-university curriculum that prepares students for the state run examinations qualifying them to join the sophomore class, Research and Higher Education are being undertaken by about 20 Universities. One is a state supported national university (a public autonomous administration under the tutelage of the Ministry of Culture and Higher Education) while the rest are private Universities. Since 1991, development of higher education has been progressing steadily, and Lebanon is quickly regaining its status as an educational center in the region, and many of the Lebanese universities, especially the American University of Beirut and the Beirut Arab University, continue to attract a considerable number of students from the surrounding countries.

One of the major higher fields of education, and accordingly professional fields through which Lebanon was serving the region, is the field of engineering. Lebanon's higher education institutions have constituted in the past, and still do, a rich source of fresh engineers for the Gulf Region. Consequently, higher education curricula are conceptualized taking into consideration regional as well as local needs of the profession. Recently, some of the Universities in Lebanon started seeking international recognition, either through gaining accreditation from international boards, such as the American University of Beirut working in close cooperation with MIT to get accredited by the ABET, or through strengthening their research and educational relations with internationally recognized institutions, such as the Beirut Arab University and the Lebanese American University. Presently, curricula are being continuously revised and modified, and the process of recruiting internationally recognized engineering scientists of Lebanese origin, as well of other nationalities, is being reactivated.

On a different level, measures are being adopted to promote research and encourage cooperation between the engineering community and the society at large:
- Universities are actively establishing close and extensive cooperation with international research institutions and laboratories;
- The process of formulating proposals for attracting internationally oriented research funds in the fields of information technology and environmental sciences has attained a considerable success;
- The establishment of the Lebanese Academic and Research Network (LARN) is almost at its final stages, a step that will enhance and facilitate local research and academic cooperation among the different institutions;
- A board consisting of representatives of the engineering faculties of most of the universities as well as the industry has been established in coordination with the National Council for Scientific Research (NCSR) to foster cooperation between the engineering 'community' and the industry.

The above, and other important developments in the Engineering Education, are being undertaken and will have a decisive positive impact on the engineering profession on the local and regional levels. The aim of this paper is to shed light on the engineering education in Lebanon, its history, its present status and future perspective.

## 1. THE HISTORY OF HIGHER EDUCATION IN LEBANON

In order to have a wide and comprehensive view about the Engineering Education in Lebanon, its historical accomplishments and its future perspectives, it is very important to review the history of the Higher Education in Lebanon, which could be divided into four phases [3]:

### *1.1. Phase I (1850-1950)*

This phase extended for almost one century. It was characterized by the domination of foreign institutions on higher education through missionaries. Though spreading education was the declared aim of these institutions, yet, gaining sociopolitical influence represented another important aim. The French sought influence through Catholic missionaries, while the British and Americans through Protestant missionaries. A struggle between the two was even originated for the sake of supremacy and this was obvious from the words of the Head of the Jesuit Fathers in a letter addressed to the French Chancellor in 1844 where he attributed the success of their mission to the failure of the Protestant missions [5]. A similar viewpoint was, as well, adopted by the Protestants personified by the first president of the American University of Beirut, Dr. Daniel Bliss [14].

*Saint Joseph University (USJ)*
In 1834, the Jesuit missionaries established an institute of theology for preparing priests and clergymen. In 1875, these missionaries have transformed this institute officially into the Saint Joseph University with its first college of Theology. In 1881, the university was officially licensed by Pope Leon the Thirteenth to issue the License and Doctorate degrees in Philosophy and Theology. In 1913, the Faculty of Engineering was established under the name "French School of Engineering". In 1936, the university was academically tied up to the University of Leon in France [10].

*American University of Beirut (AUB)*
In 1849, the American Protestant missionaries established an institute of theology for preparing priests and preaching Protestantism. This institute used to be a regional station for preachers coming from the United States to study Arabic and thus become abler to communicate with, and preach the natives. In 1866 the American University of Beirut was established under the name of the Evangelical Syrian College with colleges of Arts and Sciences. A number of colleges have followed in the fields of Medicine, Pharmacology, Trade, Nursing and Dentistry. The interesting thing about this college, at that stage, is that teaching was in Arabic [15]. Due to this, the Arabic library was enriched with scientific books translated into Arabic by the missionaries. The decision to convert into English was taken in 1884, after almost 20 years, by the Board of Trustees in New York due to a crisis that took

place in 1882 [8]. In 1920 the college was given the name of the American University of Beirut.

*The American Junior College for Women (A.J.C.W)*
The Council of Protestant Missionaries established this Institution in 1924. This college was initially a school under the name the American School for Girls. In 1950, its name was changed to Beirut College for Women (BCW).

*The Higher School of Arts and the Center for Mathematical and Physical Studies and Research*
The French Foreign Ministry established both of these institutions. The first was established in 1944 and the second in the year 1945. Both were put under the academic supervision of the University of Leon. In mid seventies both institutions were closed.

*The Lebanese Academy for Arts (ALBA)*
This institution was established by private and personal efforts in 1937. The educational system of this institute was based on combining the French and American systems and the teaching language was French. At the beginning, programs were restricted to Music and Painting but later on, programs in Architecture and Languages were added.

*The Near East School of Theology*
The Council of Protestant Missionaries established this institution in the 1883. Initially, it occupied a building inside the American University of Beirut (AUB) Campus, but became independent in 1932. This institution used to prepare missionaries for the Middle East and Africa.

In addition to the fact that this was a phase of foreign dominance on Higher Education, we could add that this phase has impregnated the Lebanese Educated Elite with either the Latin Culture through the French Educational Model, or the Anglo-Saxon Culture through the American Educational Model. This phase, as well, has framed the Higher Educational System in Lebanon for years to come.

*1.2. Phase II (1950-1975)*

This phase has witnessed the initiation of national universities, a state operated university and other private universities. The expansion of already established private universities took place, as well, during this phase.

*The Lebanese University*
This is the only state operated university that was established in 1951 [17]. The first College was that of Education and in 1959 the Colleges of Arts, Human Rights,

Sciences, Law, Political Sciences and Social Sciences were opened. In 1967, the Colleges of Business Administration, Journalism and Fine Arts were added to the university curriculum. In 1974, the College of Engineering was established on paper but did not start functioning until 1980.

*The Beirut Arab University*
This institution was established in 1960 [2] by the local Lebanese society of "AL BIRR WAL-IJSAN" in cooperation with the Egyptian government and under the direct supervision of the State University of Alexandria. The first Colleges were those of Literature, Law, Trade and Architecture. The Faculty of Engineering was established in 1975 starting with the Civil and Electrical Engineering programs. This university is the only institution in Lebanon with a distant learning program normally joined by students from almost all Arab countries. However, students have to sit for a final examination at the end of each year at the University Campus in Beirut.

*The Holy Spirit University – KASLIK (USEK)*
The Lebanese Christian Maronite Order established this institution in 1950. The first Colleges were those of Theology, Philosophy, Humanities, Literature and Law. The government in 1962 officially recognized it. The opening of the Colleges of Trade and Agriculture followed in 1966 and 1974 [13].

*The Beirut College for Women (BCW)*
In 1955, this institution was granted the privilege to issue university degrees from the Board of Trustees of the State University of New York. In 1973, its name was changed to the Beirut University College (BUC).

*The Saint- Joseph University*
This first generation university has witnessed expansion and modernization of its infrastructure under the influence of the increasing Lebanese and Arabian Gulf markets, and due to the increased competition with other existing universities. During this phase as well, the name of the "French Engineering School" was changed to the "Higher Institute for Engineering in Beirut (ESIB)", and years of study were increased to 5 years.

*The American University of Beirut (AUB)*
Similarly, this first generation university has witnessed expansion and modernization under the influence of local and regional job market needs and also due to the increased competition with other universities. In the year 1951, the School of Engineering was established followed in 1952 and 1954 with the Schools of Agriculture and Public Health. During this phase, the university was integrated in the American AID program [11], which was established by the US government to train students from the developing and under-developed countries.

*Haikazian University College*
This university was established by the Union of Evangelical Armenian Churches for the Near East in 1955, with the primary aim of educating the Armenian Youth taking into consideration in its Curricula the Armenian Language, History and Culture.

In summary we can say that this phase has witnessed the founding of several universities, one state university and the expansion and modernization of already established universities. In addition, the acceleration in the participation rate in higher education has moved Lebanon to a mass educational provider.

## 1.3. Phase III (1975-1989)

The phase of the Lebanese civil war, the fragmentation of the public administration and the emergence of demarcation lines dividing the country into areas controlled, except in rare cases, by non-governmental fighting forces.

*The Lebanese University*
The national university has witnessed devastating destruction in its infrastructure, and under the pressing war situation and due to the dividing demarcation lines, the central government decided to build different branches in the different areas for the aim of providing education for the population under siege by the different paramilitary factions. Because of the chaotic war situation and immigration of most qualified faculty members, the university has faced a considerable deterioration in the qualifications of the recruited faculty members and accordingly in the quality of education.

*Saint-Joseph University*
The civil war has forced the administration of this institution to adopt a new and more open vision taking into consideration the importance of the multicultural environment of the Lebanese society. And as with other sister institutions during this war phase, the administration has decided to open branches in different Lebanese territories [7].

*The American University of Beirut*
During the war years of 1976/77, the administration decided to open another branch for the university in the divided Lebanon. In 1984, its president Dr. Malcolm Kerr was assassinated and most of the foreign academic and administrative staff departed. Many educational programs were closed and all Ph.D. programs were suspended.

*The Beirut University College (BUC)*
Again, and due to the war, the university opened its first branch in 1978 in the city of Byblos to the north of Beirut, and in 1989 a second branch in the city of Saida to the south of Beirut.

*The Notre Dame University (NDU)*
The Lebanese Christian Maronite Order of the Holy Virgin established this Catholic institution in 1978 after the American Model of higher education. It started with the schools of business Administration and Computer Science. In 1979, the engineering program was started with 2 preparatory years in Lebanon and the rest in the USA. The government in 1987 recognized the university.

*The Beirut Arab University (BAU)*
In 1982, and due to the war situation, the university was obliged to open special final exams branches in several Arab countries for those following the distant learning program. In 1987, the School of Pharmacy was established.

*The Balamand University (BU)*
The Greek Orthodox Church founded this university formally in 1988. This institution has included two already existing institutes, the Institute of Saint John for Theology originated in 1970 and the Institute of Fine Arts originated in 1937. Two colleges were established with the opening ceremony of the university, that of literature and humanistic sciences.

In summary, this phase has witnessed the branching of some universities under the influence of the civil war and the emergence of the dividing demarcation lines. On the other hand, new universities were founded and other expanded. Foreign and Native faculty have immigrated because of the war situation, which resulted in the deterioration of the standards of higher education.

*1.4. Phase IV (since 1989)*

This phase has started in 1989 after what is known by "The Al Taif Accord", which was reached among the different Lebanese fighting factions and which took place in the city of Al Taif in the Kingdom of Saudi Arabia and put an end to the civil war.

*The American University of Beirut (AUB)*
The university was reunified in 1990 by a decision taken by the Board of Trustees in New York. In 1998, the Center of Advanced Mathematical Studies was opened aiming at supporting research in the different scientific and technological fields.

*The Beirut University College (BUC)*
This College has been officially recognized as a university in 1992 after it was reconstructed and its name was changed to the Lebanese American University (LAU). In 1996 the Faculty of Engineering was opened following the American model of education.

*The Notre Dame University (NDU)*
Its Faculty of Engineering was established in 1994, and in 1998 the university has moved to a new and modern campus.

*The Beirut Arab University (BAU)*
In 1995, the Faculties of Medicine and Dentistry were established and the Department of Mechanical Engineering was added to the Faculty of Engineering.

*AL Balamand University (BU)*
In 1992, the College of Applied Sciences was opened, and in 1993 the Faculty of Engineering was established and operated on a credit-hour semester system. In 1995, the College of Public Health was opened.

*The Islamic University*
The Higher Islamic Council of the Shiite with colleges of Theology, Arts, Sciences and Medical Engineering founded this university in 1994.

*The Beirut Islamic University*
This university was founded 1998 by the "Dar Al Fatwah", the Highest Islamic Institution of the Lebanese Government, and practically established in 2000 with institutes of Judiciary, Humanistic and Food Technology.

In summary, this was the phase of reunification of some of the branched universities, as well the reconstruction, the new construction and the expansion of these institutions.

## 2. ENGINEERING EDUCATION IN LEBANON

It can be seen from the above historical overview that the tradition of Engineering Education in Lebanon goes back to the year 1913, starting with the "French Engineering School" of Saint Joseph University, and the Faculty of Engineering at the American University of Beirut. These two systems have framed the Engineering Education in Lebanon.

Private and State supported initiatives such as the Lebanese Industrial Research Association (LIRA), the Lebanese Academic and Research Network (LARN) and the Lebanese Board of Creative Mechanical Design (LBCMD, have been launched in close cooperation with universities for the aim of assisting in, and promoting the modernization of engineering education.

## 3. WHY LEBANON AS A CENTER OF ENGINEERING EDUCATION IN THE MIDDLE EAST?

Several aspects are to be considered, which support this allegation.
The Scientific and Educational Aspects:
- The Historical Tradition of Engineering Education since 1913;
- Adopting either English or French as the teaching language;
- Enriching Engineering Curricula in all universities with Business, Cultural and Humanistic concepts;
- Modernizing Engineering Labs;
- Establishing strong links with internationally recognized universities and educational institutions from Europe and the USA;
- Recruiting highly qualified faculty members, native and foreign;
- Establishing strong research and educational networks, local and international.

Economical Aspects:
- Lebanon has a Liberal Open Market Economy and is very famous for its Banking and Service sectors;
- The Lebanese Currency is stable (average exchange rate for the US Dollar for the last 4 years is about 1500 Lebanese Pounds);
- Opportunities for Promotion and Advancement are quickly improving;
- New partnerships are being strongly forged between Lebanon and Europe through various programs (PROMOS and MEDA);
- Lebanon is still a major source of the fresh and experienced engineers to the Arab Gulf States.

Historical Aspects are:
- Lebanon's History is very rich with Historical, Cultural and Scientific events that have influenced regional and international events;
- Lebanon played a major role in the Islamic and Arab Renaissance;
- Lebanon, as well, played a major role in promoting Higher Education in the Region.

Cultural Aspects:
- Lebanon is a center of various schools of thought (social and political);
- Lebanon is a multicultural open society.

Political Aspects
- Lebanon has a stable democratic parliamentary system with freedom of opinion and press.

## 4. MEASURES TO BE TAKEN TO FOSTER THIS ROLE

The following steps are to be undertaken in cooperation with International Universities and Research Centers in order to strengthen and promote this role:
- Establish common programs for the exchange of Faculty Members on semester and yearly basis;
- Establish joint Masters and PhD programs;
- Establish common research projects of mutual benefit;
- Start Regional Continuing Educational Center for Engineers;
- Establish Technology Transfer Centers with the duty of transferring technology from the highly industrialized countries to the region in a way that befits the requirements and needs of Regional Countries;
- Conduct Industrial Training Programs for Students.

*Table 1. Universities with engineering programs sorted according to the date of initiation of the program*

| Name | Abbreviation | Diploma Program | Bachelor Degree Program | Masters Degree Program | PhD Program | Initiation Year |
|---|---|---|---|---|---|---|
| St. Joseph University | USJ | * | | | | 1913 |
| American University of Beirut | AUB | | * | * | | 1951 |
| Beirut Arab University | BAU | | * | * | | 1975 |
| Lebanese University | LU | * | | | | 1980 |
| Balamand University | BU | | * | | | 1993 |
| Notre Dame University | NDU | | * | | | 1994 |
| Lebanese American University | LAU | | * | | | 1996 |

## 5. REFERENCES

1. Al Bustani, F.A., "The University in the Arab World: Birth and Development", The Journal of Research, Nr. 2, 1955, p. 191-216 (in Arabic).
2. Hallak H. "The Beirut Arab University, 40 years", Beirut Arab University 2000 (in Arabic).
3. Alamin Adnan (Editor), "Higher Education in Lebanon", The Lebanese Institution for Educational Sciences, 1997 (in Arabic).
4. Alamin Adnan, "Education in Lebanon", Dar Al Jadid, Beirut, 1994 (in Arabic).
5. Atrissy, Talal; The JESUITS Missions, Beirut The International Publishing Agency, 1987, p. 69 (in Arabic).
6. Bashur Munir, "The Deterioration of the Educational System in Lebanon", The Arab Future, Nr. 111, May 1988 (in Arabic).
7. Ducruert, Jean S.J.: L'Université et la Cite, Beyrouth", Université Saint-Joseph, 1955, pp. 301-304.
8. Jeha Chafic, The Crisis of 1882, The Book of the Holiday 1866-1966, The American University of Beirut, 1967, p. 338 (in Arabic).
9. Khazen M.: "Jounieh through History", p. 335, 1982 (in Arabic).
10. Ristelhueber R., "Tradition françaises au Liban", Paris, 1918, p. 279.
11. Penrose S.: "The Yearly Report", American Univerisity of Beirut, 1950/1951.
12. Salem, Joseph, "The Democracy of Education, an aim to struggle for in Lebanon", The Journal of The Way, Feb. 1980 (in Arabic).
13. Thabet, John (Reverend): "The Holy Spirit University", Kaslik, 1985 (in Arabic).
14. The Reminiscences of Daniel Bliss, New York, N.Y., Fleming H. Revell Company, 1920, p. 168.
15. Tibawi A.L., "The Gensis and Early History of the Syrian Protestant College", Festival Book 1866-1966, American University of Beirut 1967, p. 223.
16. Université Saint-Joseph: "Les JESUITE en Syrie 1831-1931", Paris, les éditions Dillen, 1931, p. 8.
17. Ziadeh S.: "The Lebanese University, Guide to Faculties", 1986 (in Arabic).

*Saleh M. Chehade*
*Beirut Arab University*
*Department of Mechanical Engineering*
*Beirut, Lebanon*

S. CHANG

# SCIENCE AND ENGINEERING EDUCATION IN KOREA

**Abstract.** The status of Science and Engineering Education in Korea is presented. Due to the rapid growth in manufacturing industries, the need for more engineers has been increased in the seventies through nineties. About 24% of all bachelors degrees awarded in Korea are engineering degrees. The number of bachelor degrees in science awarded in Korea is slightly less than Japan. But the number of graduate degrees in science and engineering is much less than Japan, Germany and the United States.

The most important issue, however, is not in the quantity but in quality of the graduates. To insure quality of engineering education, the Accreditation Board for Engineering Education of Korea (ABEEK) was chartered in 1999. The Korea Council for University Education (KCUE) has also conducted University and Departmental evaluations since early nineties.

Top ten universities in Korea take most of the research fund awarded by the Government and industries. The most difficult problem facing science and engineering education in Korea is lack of funds compared to the advanced countries. Education expenses for Korean science and engineering education are far less than the G7 countries.

## 1. INTRODUCTION

The history of science and engineering education in Korea is not very long. Ten engineering students in college level graduated in 1918 after three years study, and four students in mathematics and physics graduated in 1919 at another institution.

The first University level graduates received bachelors degrees in 1943; one student in physics, four in mechanical engineering, two in metallurgical engineering, five in electrical engineering and two in chemical engineering.

Education in all levels were severely affected during the Korean War 1950-1953. Buildings and laboratories were damaged and destroyed. After the cease fire in July 1953, many professors from Seoul National University went to University of Minnesota for study and fact-finding mission. Science and engineering departments of Seoul National University received some laboratory equipment through AID plan.

The economic progress in the sixties and seventies stimulated engineering and science education at universities.

We can proudly say that engineering and science graduates from Korean universities have contributed tremendously in Korean economic developments in the last thirty years.

Korea is No.2 in shipbuilding and semiconductor memory manufacturing, No.6 in steel and automobile manufacturing in the World.

We now have nuclear reactors of our own design and a third generation 2 GeV Synchrotron.

## 2. GROWTH IN SCIENCE AND ENGINEERING EDUCATION

There were only 15 engineering colleges in early sixties in Korea. Now there are 83 engineering colleges and also over 150 junior colleges. The growth in science education is less but similarly expanded.

The number of graduates in sciences for Korea, Japan, United States, Germany and United Kingdom is shown in Table 1. Similar statistics for engineering graduates are shown in Table 2.

From table 1 and table 2, the following observations are possible:
1. Korea produces too many bachelors degrees in sciences. The population of Korea is about one third of Japan, but the output is almost same.
2. Ph.D. output in science in Korea is relatively small, but the demand is not increasing fast enough.
3. Bachelors in engineering in Korea seem to be enough compared to Japan and United States, but need more output in electrical, computer and mechanical engineering.
4. The Ministry of Education has authorized too many engineering schools in the last 10 years, so the output will be further increased.
5. The objective of science and engineering education in Korea should be increasing the quality not quantity. 75% of universities are private depending heavily on the tuition income and Government support is minimal.

## 3. GRADUATE EDUCATION IN ENGINEERING

There are 236,082 undergraduate students in 83 engineering schools in Korea as of March 1998. 77 schools have masters programs with 20,735 students and 64 schools have Ph.D. programs with 12,175 students. But the number of faculty members is only 7,293 which gives student to faculty ratio of 45.

Only 13 engineering schools have more than 200 doctoral students. 40 schools have less than 100 doctoral students. This means that there are too many small uncompetitive Ph.D. programs.

When a university is heavily graduate education oriented, it should reduce undergraduate output and predominantly undergraduate-oriented schools should not try to award too many graduate degrees.

The following graph shows that engineering schools in the United States follow this principle very well but Korean counterparts do not. Most engineering schools in the United States are located left to the curve which is called the stress line. If a school is located right to the curve, it means that professors get stress to supervise many undergraduate students and graduate students.

## 4. GOVERNMENT SUPPORT FOR HIGHER EDUCATION

The total education expense for each engineering student per year in Korea is about $5,000 including tuition payment and government support. This shows how poorly the science and engineering education is funded by the Korean Government.

Table 1. Science graduates in Korea, Japan, United States, Germany and United Kingdom

| | BS | | | | | MS | | | | | Ph.D | | | | |
|---|---|---|---|---|---|---|---|---|---|---|---|---|---|---|---|
| | Korea | Japan | USA | Germany | UK | Korea | Japan | USA | Germany | UK | Korea | Japan | USA | Germany | UK |
| Mathematics Statistics | 4,183 | 5,103 | 13,723 | 135 | 3,939 | 360 | 975 | 4,181 | 2,096 | 529 | 93 | 144 | 1,226 | 430 | 302 |
| Physics | 2,754 | 4,020 | 3,992 | 77 | 2,320 | 417 | 1,361 | 1,817 | 3,466 | 315 | 112 | 300 | 4,124 | 1,586 | 558 |
| Chemistry | 3,156 | 3,669 | 9,722 | 394 | 3,393 | 651 | 1,159 | 2,099 | 2,981 | 427 | 124 | 175 | 2,273 | 2,564 | 927 |
| Life Sciences | 3,889 | 2,011 | 55,984 | 167 | 16,885 | 673 | 832 | 5,393 | 2,936 | 1,947 | 153 | 194 | 4,625 | 1,693 | 1,674 |
| Geology and Earth Science | 448 | 831 | 3,118 | 11 | 1,221 | 70 | 484 | 993 | 924 | 311 | 22 | 98 | 398 | 315 | 154 |
| Computer Science | 3,735 | | 24,404 | 3,042 | 9,991 | 517 | | 10,336 | 3,561 | 2,763 | 92 | | 884 | 355 | 263 |
| Others | 2,783 | 2,855 | | 35 | 3,503 | 267 | 456 | | 3,649 | 409 | | 136 | | 389 | 198 |
| Total | 20,948 | 18,489 | 110,943 | 3,861 | 41,252 | 2,955 | 5,267 | 24,809 | 19,613 | 6,701 | 596 | 1,047 | 13,530 | 7,332 | 4,076 |

Sources :

Korea : Statistical Yearbook of Education 1998

Japan : Report on Schools 1998, Ministry of Education
Computer Science included in electrical and Communication engineering

U S A : Digest of Education Statistics 1997.

Germany : Statistisches Jahrbuch 1999
Diplom Ingenieurs are included in MS
Graduates of Fachhochschule in B.S.

UK : Higher Education Statistics Agency 1997/98

France : Doctors 6,290, DEA&DESS 16,149, Maîtrises 24,544, Licences 32,475

Table 2. Engineering graduates in Korea, Japan, USA, Germany and United Kingdom

| | B.S | | | | | M.S | | | | | Ph.D. | | | | |
|---|---|---|---|---|---|---|---|---|---|---|---|---|---|---|---|
| | Korea | Japan | USA | Germany | U.K | Korea | Japan | USA | Germany | U.K | Korea | Japan | USA | Germany | U.K |
| Aeronautical | 345 | 645 | 1,771 | | 657 | 92 | 173 | 821 | | 148 | 9 | 28 | 229 | | 28 |
| Agricultural | | | 591 | | | 6 | | 160 | | | | | 86 | | |
| Architecture | 4,331 | | 558 | 3,984 | 7,273 | 630 | | 25 | 2,895 | 1,235 | 80 | | 4 | 63 | 134 |
| Biomedical | | | 913 | | 129 | | | 480 | | 113 | | | 175 | | 16 |
| Chemical | 3,911 | 11,303 | 5,901 | | 1,116 | 636 | 2,867 | 1,085 | | 227 | 106 | 305 | 571 | | 191 |
| Civil | 3,157 | 19,868 | 9,927 | 4,533 | 3,312 | 686 | 3,097 | 4,077 | 2,516 | 708 | 103 | 244 | 625 | 250 | 225 |
| Computer | 3,802 | | 2,345 | | | 393 | | 1,040 | | | 58 | | 140 | | |
| Electrical | 9,590 | 31,773 | 14,929 | 7,131 | 4,867 | 2,069 | 6,826 | 7,693 | 4,564 | 1,162 | 345 | 552 | 1,543 | 559 | 374 |
| Environmental | 1,670 | | 641 | | 2,755 | 442 | | 1,056 | | 731 | 35 | 21 | 45 | | 111 |
| Industrial | 2,859 | 5,189 | 3,147 | | 2,332 | 908 | 247 | 2,061 | | 619 | 49 | 21 | 296 | | 60 |
| Materials Science | 2,860 | 687 | 769 | 86 | 1,321 | 585 | 147 | 671 | 209 | 272 | 118 | 29 | 455 | 122 | 276 |
| Ceramic | 149 | | 191 | | 22 | | | 88 | | 6 | | | 57 | | 8 |
| Mechanical | 6,018 | 20,676 | 14,794 | 11,995 | 3,913 | 1,161 | 3,874 | 4,213 | 6,801 | 550 | 148 | 186 | 890 | 1,218 | 254 |
| Naval Architecture | 542 | 249 | 278 | | | 81 | 75 | 30 | | | 15 | 13 | 6 | | |
| Mining | 397 | 197 | 121 | | 123 | 61 | 90 | 72 | | 86 | 10 | 13 | 15 | | 29 |
| Nuclear | 158 | 411 | 247 | | | 43 | 210 | 264 | | | 4 | 27 | 111 | | |
| Petroleum | | | 303 | | | | | 200 | | | | | 51 | | |
| Others | 7,572 | 10,942 | 4,916 | 1,070 | 4,886 | 1,471 | 5,731 | 4,517 | 315 | 1,187 | 114 | 976 | 811 | 80 | 379 |
| Total | 47,361 | 101,940 | 62,342 | 28,799 | 32,706 | 9,264 | 23,337 | 28,553 | 17,300 | 7,044 | 1,194 | 2,394 | 6,110 | 2,292 | 2,085 |

Sources : Same as Table 1
Korea : B.E.T 10,069 not included
Japan : number of Ph.D. include only the new system, the old system without any course requirements produces approximately same number.
U.S.A. : B.E.T 15,812 not included

# SCIENCE AND ENGINEERING EDUCATION IN KOREA

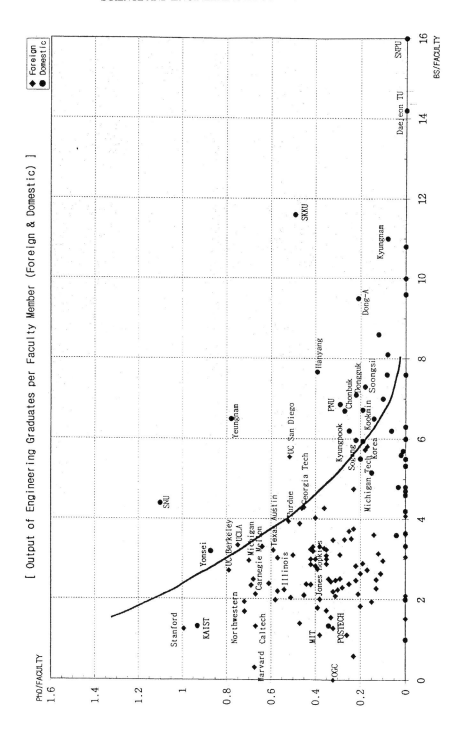

*Table 3. Government support budget for higher education in several countries*

|         | Government Budget in Higher Education | Students  | Budget per Student | |
|---------|---------------------------------------|-----------|-------------------|----------|
| Swiss   | SFr 4,038 million (1998)              | 91,800    | SFr 43,987        | ($29,324) |
| Germany | DM 67,323 million (1998)              | 1,822,898 | DM 36,932         | ($18,842) |
| Sweden  | SEK 31,500 million (1995)             | 269,800   | SEK 116,753       | ($15,052) |
| Japan   | ¥3,342,871 million(1994)              | 2,533,342 | ¥1,319,549        | ($11,333) |
| U.K.    | £5,314 million (1993)                 | 967,942   | £5,490            | ($9,043) |
| U.S.A.  | $ 68,985 million * (1994)             | 8,749,181 |                   | ($7,885) |
| France  | Fr 52,000 million (1995)              | 1,560,000 | Fr 33,330         | ($6,600) |
| Australia | A$ 4,649 million (1992)             | 559,365   | A$ 8,312          | ($5,569) |
| Korea   | ₩ 1,700,541 million (1997)            | 1,825,020 | ₩ 931,793 + $4,200 Student tuition | ($776) |

\* Federal and State Governments

## 5. HOPES IN THE FUTURE

The Ministry of Education is funding $166 million annually for graduate education in science and engineering particulary in information Technology, Biotechnology and other areas (Mechanical, Materials Science, Chemical Engineering, Physics and Chemistry) in the next seven years.

This is the first such large funding for graduate education in science and engineering for Korea.

The program for science research center and engineering research center began 1991 which has contributed much for advancement of science and engineering in Korea.

Industrial research contracts usually requires immediately applicable results rather than long-term goals. It is important that industries should be persuaded to sponsor future-oriented projects as well as the short-term projects.

The Korea Council for University Education (KCUE) has conducted University and Departmental evaluations since early nineties. Due to KCUE evaluations, University education and engineering education in particular have improved significantly.

To insure quality of engineering education, the Accreditation Board for Engineering Education of Korea (ABEEK) was chartered in 1999.

In spite of general lack of funding, the quality of engineering education in Korea is expected to be improved greatly due to the recent special program initiated by the Ministry of Education and ABEEK accreditation activities.

*Sooyoung Chang*
*Pohang University of Science & Technology*
*Pohang, Korea*

# PART IV.

# EUROPEAN BACHELOR AND MASTER PROGRAMMES

R. NEGRINI

# THE NEW ENGINEERING BACHELOR PROGRAMMES IN ITALY AND AT POLITECNICO DI MILANO

**Abstract.** This paper treats the new structure of engineering studies in Italy, that starting from academic year 2000/2001 introduces bachelor programmes for all the students.
In particular, the paper considers the approach followed at Politecnico di Milano, that in six of its seven campuses organizes bachelor programmes for engineering of 16 different types, and educates more than one fifth of the Italian graduate engineers.

## 1. THE ITALIAN SCENARIO

After long discussions and debates, Italian Polytechnics and Universities are now facing the newborn structure of engineering programmes.

The previous bachelor programmes (the *Diploma* courses introduced just eight years ago) were kept on different tracks (with different subjects), separate from the historic five years Engineering *Laurea* courses. These "parallel" bachelor programmes were followed only by those students that had the feeling that the five years laurea was too long or too heavy for them.

After graduation, the "Diploma" students could ask for admission at the third year of a laurea course (i.e., with the loss of one full year of study!): this admission was very rare, and was considered a true change of vocation. Moreover, Diploma programmes were not sufficient for admission among chartered professional engineers in the Italian *Ordine degli ingegneri*.

As an obvious consequence, laurea programmes were largely preferred (see Table 1): diploma programmes were costly for the Italian universities, and not adequately productive.

*Table 1. Number of university newcomers in Italy during 1999/2000*

|  | Laurea | Diploma | Total |
|---|---|---|---|
| All programmes (incl. Engineering) | 246,131 | 32,172 | 278,303 |
| Engineering | 28,658 | 4,759 | 33,417 |

Actually, the Italian society never considered the Diploma as a respected alternative to the old honorable Laurea.

The old laurea programmes, on the other hand, had too many dropouts and too subjects far from engineering practice. Students passed too many years at the university: more than 7 years for the average graduate engineer - note that in Italy access to university is allowed after 13 years at school, and that the average beginner is 19 years old).

In some sense, the old Laurea was too complete and too heavy, to be the only real way to become an engineer. It was a true *over-education* with no alternatives.

It should be noticed that, as a matter of fact, laurea programmes in engineering were in a much better position, in comparison to many other laurea programmes of different kind, in terms of dropouts, of education to real professional need, and of employment.

The only real trouble for engineering was the average length of the studies.

A rough and global index of dropout for all Italian university programmes (including engineering, and including Laurea and Diploma programmes) is given in Table 2; for engineering only, the situation was better (consider, e.g., that Politecnico di Milano graduated roughly half of its newcomers to engineering laurea programmes).

A real change was considered necessary by the Government and by many academic authorities.

*Table 2. Number of students of all kinds of university programmes in Italy*

| Newcomers (Laurea or Diploma) in 1992/93 | Graduated (Laurea or Diploma) during 1997/98 |
|---|---|
| 362,006 | 131,927 |

Another very strong reason put forward in favor of a change was the desire for a more integrated European structure for university studies (the well known *Joint Declaration* (1998) was cardinal for the start of this renovation process in Italy).

## 2. THE NEW STRUCTURE OF ENGINEERING STUDIES IN ITALY

Now, all the engineering students will take a new bachelor programme of three years (the new, lighter *Laurea*), and then continue with a two years master programme (the *Laurea Specialistica*, equivalent to the old Laurea).

The new bachelor programme must provide a good professional knowledge in a more reasonable amount of years.

All the Universities and Polytechnics give grades that have the same legal value (e.g. for admission to the professional Order of chartered engineers in Italy, or to compete for a job position in a public body).

The new organization maintains this "legal uniformity", but a large debate is now open over this point, that many consider an obstacle to the true competition among universities and an undue advantage for universities with lesser reputation and weaker educational capabilities. Indeed, it contrasts with the larger freedom now given to the universities, that now have more responsibility in the definition of programmes and can follow quite different paths towards degrees that have the same legal value.

Two general directives of *Ministero dell'Università e della Ricerca scientifica e tecnologica* (Italian State Ministry) set the broadly generic frame for all Italian university bachelor programmes (see *Regolamento* (2000) and *Decreto ministeriale* (2000)), leaving great freedom to the Universities. In fact, the local "Programme Council of Professors" determines the rules of a local programme, and the Academic Senate of the University approves them.

For an easier international integration, these directives dictate the generalized adoption of *credits:* one credit corresponds to 25 hours of work for an "average" student (an engineer would probably prefer more practical definitions of units of measure).

Each year of study must correspond to 60 credits. Credits are gained not only through exams, but also through final work/projects, extramural education, abilities in foreign languages, etc.

The new structure of engineering studies is depicted in Fig. 1, following the so-called "3+2 years schema" for graduation, but note that universities can offer *part-time programmes*, of longer duration, in order to help students that must work to help themselves.

At a first glance this structure seems quite similar to a classic "bachelor plus master" one, well known in other European and American traditions.

The most important difference is that Laurea and Laurea specialistica of the same kind (e.g., both of Computer Engineering) are linked together in an implicit manner: to obtain a Laurea Specialistica, the number of credits in subjects of a predefined kind (e.g., of characterizing disciplines) cannot be gained in two years only: credits from the bachelor programme are also needed. Thus, passages from bachelor programmes to graduate programmes of different kind (e.g., from Mathematics to Computer engineering) become quite difficult.

Admission to engineering bachelor programmes must not have *numerus clausus* constraints: Universities can autonomously limit the access only to special or new programmes. All the students, after 13 years at school, can choose engineering, independently from professional, technical or humanistic schools.

In any case, Universities are free to impose compulsory entrance tests to every programme, and can require that students that do not pass these tests take some extra subjects during their first year (e.g., school-level mathematics), if possible at the beginning, before having full access to standard subjects.

Programmes are constrained by the State rules only through the definition of "classes" of programmes.

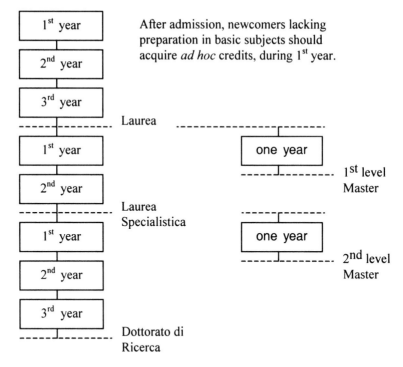

*Figure 1. The new structure for engineering courses, based upon 3 years bachelor programmes (Laurea) followed by 2 years graduate programmes (Laurea Specialistica) and then by 3 years PHD courses (Dottorato di Ricerca).*
*Shorter "Master" programmes of one year can be offered after Laurea or after Laurea Specialistica.*

Inside each class (e.g., Information Engineering class) all the programmes (e.g., Computer Engineering, or Telecommunications Engineering) must require more than a fixed amount of credits (typically, 108 out of 180) for subjects chosen in given disciplinary sectors.

For example, Telecommunications engineering requires, at least:

- 27 credits of basic mathematics, physics, chemistry, and informatics;
- 36 credits of disciplines characterizing the course (electromagnetic fields, telecommunications);
- 18 credits of related complementary disciplines (electronics, informatics, automation or other engineering subjects);
- 27 credits of scientific culture, linguistic abilities, final project, etc.

Every university can choose the subjects that can satisfy these constraints (e.g., to have or not to have subjects on chemistry and physics).

The remaining 62 credits are given to subjects freely chosen by every university: in practice, this roughly states the complete freedom of choice of subjects for one year of study.

Among the 42 classes defined at National level, the following ones are naturally adapt to engineering programmes:

1. Class of Sciences of Architecture and Building Engineering;
2. Class of Civil and Environmental Engineering;
3. Class of Industrial Engineering;
4. Class of Information Engineering.

## 3. ENGINEERING STUDIES AT POLITECNICO DI MILANO

Politecnico di Milano, that graduates more than one fifth of the Italian "laureate" engineers, is heavily involved in the definition of the new models of Italian undergraduate and graduate engineers, in order to improve technical education by means of smaller classes and more laboratories, with less time losses and student dropouts.

Today Politecnico di Milano offers Architecture and Engineering bachelor programmes in seven campuses. Engineering is offered in Milan (campus Leonardo and campus Bovisa), Como, Lecco, Cremona, and Piacenza (see Fig. 2). The engineering programmes are listed in Table 3.

All of them last three years and grant the new *Laurea* (e.g. *Laurea in ingegneria aerospaziale, Laurea in ingegneria biomedica,* etc.)

A three years bachelor programme can hardly prepare to a complex professional work and at the same time give a complete and solid education both in basic sciences and in theoretic fields linked to engineering.

For this reason, each bachelor programme of Politecnico is split in two paths (usually different at the third year) so that one path is devoted to a better professional education, with more subjects linked to engineering design practices and more interactions with companies and industries. The other one has a better basic education.

Both paths give a sound answer to professional needs, but only the student that follows a theoretically oriented curriculum is admitted to the Laurea Specialistica courses without "debts": the other students must take some additional theoretic subjects during their first year at graduate level. Usually, the total amount of additional credits corresponds to a semester (or even less) and depends on the particular graduate programme that the student asks for (see Fig. 3).

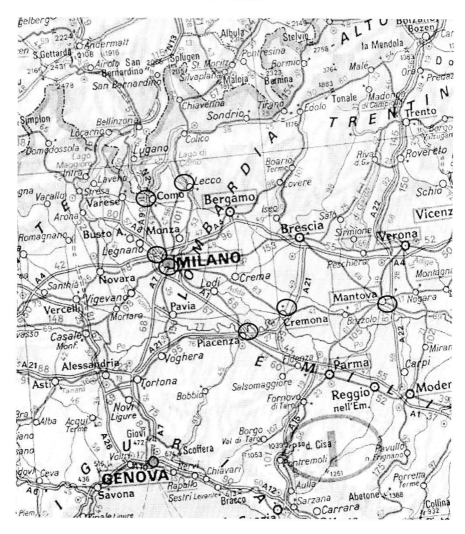

*Figure 2. Location of Politecnico di Milano seven campuses (Milan Leonardo and Milan Bovisa, Como, Lecco, Piacenza, Cremona and Mantua).*
*Ten years ago Milan Leonardo was the only, historic campus of Politecnico.*

Table 3. Location of the new bachelor engineering programmes (Laurea) at Politecnico campuses, and number of newcomers (academic year 2000/2001).
Politecnico offers Laurea specialistica courses of the same kind, plus Nuclear engineering. Moreover, a five years course in Building Engineering and Architecture is located in Lecco. Other new programmes are under consideration, both at Laurea and at Laurea Specialistica level.

| Bachelor Programmes at Politecnico (2000/2001) | Nr. of $1^{st}$ year students | Campus |
|---|---|---|
| Aerospace engineering | 368 | Milan (Bovisa) |
| Biomedical engineering | 231 | Milan (Leonardo) |
| Building engineering | 148 | Milan (Leonardo), Lecco |
| Chemical engineering | 111 | Milan (Bovisa) |
| Civil engineering | 232 | Milan (Leonardo), Lecco |
| Computer engineering | 784 | Milan (Leonardo), Como, Cremona |
| Computer engineering (On Line) | 185 | Como |
| Electrical engineering | 151 | Milan (Leonardo) |
| Electronics engineering | 237 | Milan (Leonardo) |
| Energetics engineering | 79 | Milan (Bovisa) |
| Environmental and land planning engineering | 262 | Milan (Leonardo), Como, Cremona |
| Management and production engineering | 634 | Milan (Leonardo), Como, Cremona, Lecco |
| Materials engineering | 58 | Milan (Bovisa) |
| Mechanics engineering | 588 | Milan (Bovisa), Lecco, Piacenza |
| Telecommunications engineering | 415 | Milan (Leonardo) |
| Transports engineering | 76 | Piacenza |
| *total* | *4560* | |

Of course, also the students coming to Politecnico for access to a graduate course, after a bachelor degree gained elsewhere, will be screened in the same way, in order to establish, for each single graduate student, the amount of additional credits required.

Note that this is not required by the State level rules for engineering: it is a policy autonomously decided by Politecnico.

State rules for admission to Laurea Specialistica programmes allow for *numerus clausus*, selective acceptance tests, attitude tests and so on: Universities are free to decide their own admission policies.

## 4. CHARACTERISTICS OF BACHELOR PROGRAMMES OF POLITECNICO

Although State rules do not allow for *numerus clausus* in bachelor programmes, Politecnico has a long tradition of access selection. For years, both the old Laurea and Diploma programmes had strict access limitations; then, restrictions to admission to Laurea programmes were cancelled by a State law.

Politecnico will continue to enforce an admission test, and this test will probably attribute "debts" (i.e., additional credits required in particular subjects) to bachelor students, according to the possibilities allowed by the new rules, as discussed in Section 2.

This test is based on questions that can be afforded by the majority of students, independently from their curricula at school. Answers are in closed form, so that all the thousands of tests can be examined in a computerized way in very few days.

Every candidate student, through test results, can autonomously foresee in a reasonable way the personal attitude to engineering studies, and the personal expected degree of success. Self-evaluation and self-orientation are today the main goals of this test.

Abilities in a foreign language are not tested for access: all the students must show these abilities in order to graduate, but can acquire them during their bachelor studies.

There are no standard lessons of foreign languages, or exams inside Politecnico. Students receive help (e.g., they have access to multimedia classrooms to practice with common self-instruction courses based on CD-ROMs, and some schoolteachers are sporadically called to give help) but they must exhibit some valid external proof of knowledge. For example, the TOEFL test is routinely adopted for English.

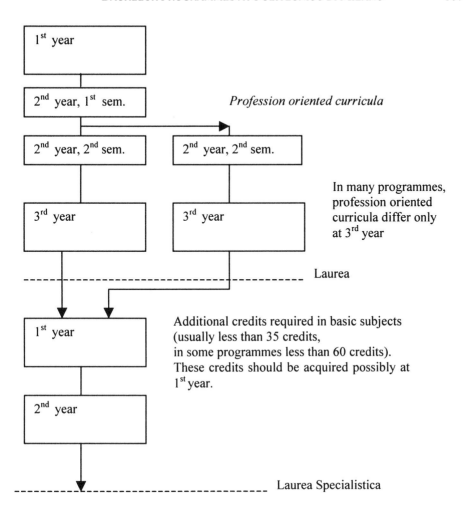

*Figure 3. At Politecnico di Milano: passage to Laurea specialistica for students coming from professionally oriented bachelor curricula.*

All the bachelor subjects are taught in Italian; on the contrary, it is easy to foresee that some subjects in the graduate programmes will be taught in English, in order to stimulate participation of students to mobility programs (as, e.g., exchange programs at European Union level, TIME double degree interuniversity agreements, Unitech interuniversity programme, and so on).

Politecnico di Milano, for its reputation in many research fields and its very large and complete offer of programmes, attracts students from all parts of Italy, and from abroad. Obviously, the most of its bachelor students comes from the Lombardy region, with roughly 9 million inhabitants.

The main obstacle found by students coming from far places is the insufficient number of student rooms and student apartments in Milan; State and Regional support are improving these conditions.

Historically, lessons in Italian universities are separate from examinations. This is generally true also for Politecnico, where the typical subject in an old laurea programme roughly corresponds to 100 hours of lessons, practical exercises and (when needed) laboratories. Examinations can be very heavy, and sometimes there is no test or partial examination during the semester. Every year the student has *seven* possible dates to pass an examination.

This organization stresses the students, and requires a good quantity of self-control and self-study capabilities. Many professors adopt partial tests, but the final examination tends to remain quite heavy.

On the basis of the experience gained in old *Diploma* programmes, where examinations were effectively distributed in smaller tests during the semester, subjects in the new bachelor programmes will be taught to smaller classes, with more practical exercises and more laboratory hours. Examinations will be distributed during the semester, with no more than two possibilities (per year and per subject) for the student to recover an examination partially failed.

Subjects will have different weights (from 5 to 12.5 credits – note that an average subject of the old Laurea corresponds to 10 credits).

Goal of thèse changes is, of course, a reduction of student dropouts, maintaining the same learning and educational qualities.

The teaching staff for undergraduate and graduate programmes is the same; full professors and associate professors of Politecnico (all with tenure) and visiting professors from other universities are mixed with experienced researchers and professionals from research institutions and companies.

Professors feel more attraction for teaching in graduate programmes for various reasons: more cultivated students, capable of autonomous work, and less burden for examinations.

Particularly useful for professors teaching at graduate level is the possibility of assigning very hard theses to the best students. In this way many professors receive a concrete help in their research activities, and cooperate with the best students as they were young and very resolute colleagues. It is even possible to find students that work for one year, full time, at their thesis, with complete personal satisfaction.

Of course, the final work for bachelor programmes cannot be of the same weight and value.

The quality of graduate and undergraduate programmes at Politecnico is scrutinized in multiple ways.

Performances of students and professors are evaluated by an internal Observatory. In particular, all the professors are surveyed by means of anonymous student questionnaires that check the quality of all didactical aspects. Results are communicated to the professor and to academic authorities.

Other internal bodies check the scientific and didactic productivity of the various groups of professors, influencing the allocation of resources to Departments and research groups.

Programmes receive financial support mainly from student fees and State revenues, and particular programmes can be supported in various ways by local entities.

For example, municipalities, provinces and chambers of commerce support their local campuses of Politecnico. This local help stimulates the flow of extra support from the Ministry.

## 5. THE ON-LINE BACHELOR PROGRAMME OF COMPUTER ENGINEERING

The most peculiar new bachelor programme is that of Computer Engineering, delivered through Internet. It is a new experience for Italy, a case of distance learning (or, better, of e-learning) based upon the net and multimedia technologies. The programme has contents very similar to those of the "in presence" Computer Engineering programmes, and is targeted to people and professionals that for whatever reason cannot follow the standard courses (typically, because of their job constraints).

This programme is based upon many years of experience of adoption of innovative technologies in didactic activities at Politecnico.

Organized in virtual classrooms, the distant students can interact through internet among themselves and with their professors, class tutors and subject tutors, both openly and privately. This on line interaction has also synchronous activities, for cooperative work done by the students and the tutors.

Lessons are substituted by multimedia presentations on CD-ROMs: globally, from 50 to 60 CD-ROMs will contain all the lessons and support material.

"In presence" activities include meetings and examinations at the Como campus; during examinations, the students also discuss tests and work assigned to them through the net, during the semester.

This programme is organized with the strong cooperation of private companies (*Somedia*, a company specialized in continuing education; *Kataweb*, internet provider that allows for easy direct connection in every part of Italy, and *Fabula*, specialized in multimedia editing).

## 6. REFERENCES

*Joint declaration on harmonisation of the architecture of the European higher education system by the four Ministers in charge for France, Germany, Italy and the United Kingdom,* Paris, the Sorbonne, May 25 1998. Available: http://www.murst.it/progprop/autonomi/sorbona/sorbgb.htm

*Regolamento in materia di autonomia didattica degli Atenei (Decreto Ministeriale* November 3 1999, nr. 509 – printed in the Italian *Gazzetta Ufficiale* nr. 2 - January 4, 2000).
New National reference frame for every didactic activity of all the Italian Universities (in Italian).
Available: http://www.murst.it/regolame/1999/adqGU.htm

*Decreto ministeriale recante determinazione delle classi delle lauree universitarie (Decreto Ministeriale* August 4 2000, printed in the Ordinary Supplement nr. 170 of the Italian *Gazzetta Ufficiale* nr. 245 - October 19, 2000).
New National reference frame for every Italian *bachelor* university programme (in Italian).
Available: http://www.murst.it/atti/decreti_area/vers1/classilauree.htm

*Politecnico di Milano* web site: http://www.polimi.it

*Roberto Negrini*
*Politecnico di Milano*
*Dipartimento di Elettronica e Informazione*
*Milan, Italy*

A. KÜNDIG

# RESTRUCTURING A UNIVERSITY LEVEL ENGINEERING CURRICULUM –

# A POSSIBLE RESPONSE TO THE BOLOGNA DECLARATION

**Abstract.** The EE department at ETH Zürich has recently undertaken a major revision of its curriculum. In the context of this revision, we have attempted to define the *profile of a university level engineer* (as opposed to a *Fachhochschule/Haute École Spécialisée engineer*). This is regarded as one of the really crucial aspects of the present reform and one of the major issues of this contribution.
The paper presents a *standard two-cycle curriculum structure*, with the first cycle (i.e. the undergraduate part) leading to a Bachelor's certificate, followed by a second cycle (i.e. the graduate part) leading to a Master's degree. In contrast to the *Bologna Declaration* [1], a Bachelor's degree in Engineering Sciences is neither seen as a professional qualification, nor is it a compulsory prerequisite for graduate admission within the same institution. Rather, we show that *a true professional qualification for a university level engineer is only attained at the Master's level*; as a consequence, the *normal university level engineering curriculum is a 4-5 year integral course leading directly to a Master's degree*. Thus, the *Bachelor's level is primarily interpreted as an intermediate mobility pivot*, i.e. an interface where the conditions for a student transfer between cooperating institutions are well defined, or where industrial practice is sandwiched between the two cycles.

## 1. INTRODUCTION

The EE department at ETH has transformed itself in the last 15 years into a real "high-tech" department, with leading-edge research in areas such as high performance communication, mechatronics, biomedical electronics, and the underlying technological basis. This has lead to a major revision of our curriculum, to become effective in 2001. Rather early in the revision process, it became clear that the introduction of an intermediate (i.e. Bachelor's) level after approx. 3 years was desirable for the following reasons:
- we want to open up our second cycle for excellent, highly motivated students from other countries, e.g. with a view to recruit outstanding PhD candidates, contributing to maintaining and raising the quality of our research.
- for our own students, we want to offer the possibility to pursue graduate studies at a renowned partner university, possibly returning as a PhD student.

In the context of this revision, we have first attempted to define the *profile of a university level engineer* (as opposed to a *Fachhochschule/Haute École Spécialisée engineer*). This is regarded as one of the really crucial aspects of the present curriculum reform; as such, it was also endorsed by a working group[1] set up by the *Conference of the Rectors of Swiss Universities (CRUS)*.

## 2. AN ATTEMPT TO DEFINE *ENGINEERING* AND *ENGINEERING SCIENCES*

### 2.1. Engineering activities and skills

It is quite clear (and undisputed) that engineering draws heavily on insights gained in the natural sciences. In fact, most definitions found in lexica stress this point, with e.g. the Oxford Dictionary [4] defining *engineering* as the "application of science to design, building and use of machines etc.". The same dictionary defines an *engineer* as a "person skilled in a branch of engineering; person who makes or is in charge of engines etc.; person who designs and constructs military works; mechanic, technician". In a very nice memorandum [11], the executive director of the US National Society of Professional Engineers states that three principal elements go along with engineering:
- engineering is linked to the "forces of nature"
- it refers to the "use or good of man"
- it implies a special knowledge and skill relating to natural or physical phenomena.

In fact, all these definitions reveal a view of engineering commonly found not only with the interested layperson, but also with people in university management and science policy – a view, however, which is still too superficial to allow us to give *university level engineering* a clear-cut profile as opposed to the profiles of other types of engineers. Evidently, the dictionary definition offers quite a range of such types, and more or less explicitly, it also suggests some rank order in the sense that (natural) scientists are "on top" – presumably since engineering is using the results of their research - way "down" to mechanics running some engine (e.g. a locomotive). In fact, ordering such activities in the way it is sketched here may be looked at from two completely different viewpoints:

(1) It is the classical (albeit too simplicistic) sequential view of the *innovation process*, where e.g. product development follows so-called applied research using itself results from basic research. Fortunately, it is more and more accepted that actual innovation processes are more complicated in the sense that e.g. feedback from the "later" stages of the process to the "early" stages of the process are rather important, e.g. the course of activities often qualified as "basic research" may well be implicitly or explicitly influenced by possible applications.

(2) It suggests some *rank order in terms of the skill (and knowledge) levels* associated with either designing, building or operating some machine or system. Although such an ordering is quite appealing at first sight, it soon becomes clear that skills and know-how required for certain engineering activities moreover

---
[1] The so-called *Engineering Sciences Task Force (ESTF)* of CRUS was chaired by the author. He thanks the members of ESTF for the many fruitful discussions and contributions.

depend on the *complexity* of the systems involved (e.g. operating a crane is less demanding than being in charge of operating a nuclear power plant, or designing a door bell is much easier than designing the next-generation microprocessor).

These considerations lead to a three-dimensional view of the different engineering disciplines as illustrated in figure 1.

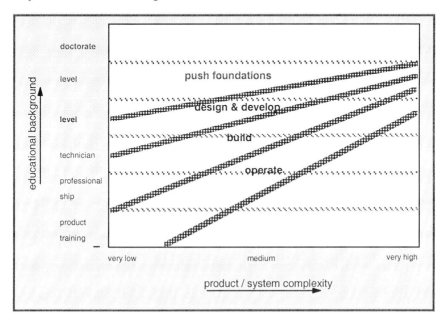

*Figure 1. Educational background required for different engineering activities and different degrees of system complexity.*

Note that the lines drawn above should not be seen as formally defining the scope of certain institutions; i.e. the figure should be looked at in a qualitative way.

## 2.2. Engineering Sciences: The sciences of synthesis

Last but not least, we would like to stress that there is a significant difference between *natural sciences* and *engineering sciences* in terms of the prevailing paradigms of the profession: With the danger of presenting a too simplicistic view, we may say that the work of the *natural scientist* aims at **analyzing** nature, whereas an *engineer* is **synthesizing** systems such that these systems fulfill certain requirements. Of course, there are nevertheless many commonalities in the sense that e.g. both use the same types of models, they have similar backgrounds in mathematics, physics and chemistry (and probably more and more in biology as well), and that they follow a scientifically based approach. Beyond that, however, the synthesis work of the engineer requires additional skills and knowledge in areas relevant for the success of the products he develops, e.g. understanding user requirements, ergonomics, economics and market characteristics, legal regulations, project management, etc. These considerations not only impact curriculum *content* (in the sense that e.g. management courses may be offered to engineering students); they will also lead us to better understand the *difference in mission between Fachhochschule and university level education* as discussed in chapter 5.

Thus, we should not use (as it is often done) the terms *science* and *engineering* to differentiate between two otherwise much related fields; rather, we must differentiate between *natural sciences* and *engineering sciences*, with engineering sciences on an equal footing with other sciences[2]. In fact, as it has been said above, an important aspect of engineering sciences is its *holistic approach* naturally taken when typical complex engineering problems are tackled. This is in stark contrast to the classical *reductionist approaches* often taken in the natural sciences. In fact, considering that the result of an engineer's work is ultimately aimed at fulfilling some needs of *people* (ranging from the actual users of the product to the shareholders), while also respecting the *environment*, the engineer's work bears considerable similarities to other professions dealing with people[3] (and their needs, desires, dreams, conflicts and also their shortcomings as humble human beings) such as e.g. medical doctors or the specialist in constitutional law. Thus, in addition to understanding the natural sciences as far as these are relevant as a source of new technologies and that they impose physical boundary conditions, a well-rounded engineer should also be knowledgeable in scientific fields related to *people* and the *environment*. In fact, it is often underestimated how far so-called straightforward engineering solutions are heavily influenced by non-technical paradigms (as e.g. bookkeeping practices or, on a larger scale, exemplified by the rules guiding engineers in the former monopolistic telecommunication sector), and how far, on the other hand, a fresh look at and possibly a change of such boundary conditions may lead to true innovations. Again, the more complex the systems involved are, the more familiar should a designer of innovative systems be with related scientific fields if a real breakthrough is to be achieved (see figure 1 again).

We argue that a university level environment provides a much better fertile soil for such proper *engineering **sciences*** than those types of institutions (e.g. Fachhochschulen) where disciplines outside the technical fields are just taught on a pragmatic "tool level". This reasoning is intimately linked with the finding that many groundbreaking innovations occur within the overlapping "grey zones" of traditional disciplines.

---

[2] A nice quotation in [11] illuminates this point from another angle: " It is easier to distinguish between the 'scientific function' and the 'engineering function' than to distinguish between the man who should be called a scientist and who should be termed an engineer. Many men perform both functions, and do it very well...".

[3] Conversely, it might be worthwhile to apply the approach taken here to other professional fields. Of course, specific attributes (such as the ones used in figure 1) for these fields would have to be found, replacing those used here for engineering.

## 3. ENGINEERING EDUCATION IN SWITZERLAND

### 3.1. Types of institutions offering engineering education

In the preceding discussion (i.e. visualized in figure 1), four different levels of tertiary education have been considered. In Switzerland, these four levels are represented as shown in table 1.

Table 1. Overview of Swiss institutions offering tertiary engineering education

*) Formerly ETS = École Technique Supérieure and HTL = Höhere Technische Lehranstalt

|  | type of institution | | |
|---|---|---|---|
| level | English term | French term | German term |
| technician (certificate) | Technical School | École Technique | Technikerschule TS |
| (college) graduate | University of Applied Science | Haute École Spécialisée HES*) | Fachhochschule FH*) |
| university level diploma | University, Federal Institute of Technology | Université, EPF | Universität, ETH |
| doctorate | University, Federal Institute of Technology | Université, EPF | Universität, ETH |

In addition, some institutions offer special programmes leading to various intermediate levels. E.g. both ETHs and many *Fachhochschulen* offer "Nachdiplomstudien" (post-diploma programmes); as a rule, these are not necessarily equivalent to stepping from one of the levels shown in table 2 to the next higher level.

If engineering is defined in a narrow sense, the (cantonal) universities would have to be excluded from the table above. However, in some specific fields, *curricula more or less close to engineering* are offered by universities such as *informatics* or *applied physics and mathematics*. Another "grey zone" is represented by *architecture*, where creative and even artistic skills are a crucial aspect of the profession, even more important than in engineering as discussed in section 2.2. Also, in some cases, departments offer programmes leading to graduates with both an "engineering sciences flavour" and a "natural sciences flavour". A typical example of such a programme is the curriculum in *Environmental Sciences (Umweltnaturwissenschaften)* at ETH Zürich[4]. This leads us also to another important observation: Figure 1 and the arguments put forward in chapters 2 and 5 may well be applied to disciplines with an "engineering sciences flavour" if the

---
[4] Incidentally, this example reveals also the fact that *science* is often understood to mean *natural sciences* (see 3.3.1).

*product development* focus of the narrow-sense engineer is replaced by a *system or strategy development* focus. For example, the efforts made in the 1950's and 1960's to master the growing problems with water pollution resulted in a new *strategy* to cope with a very complex, multidisciplinary (man-made) problem. Last but not least, such new strategies may ultimately trigger the development of a host of new products such as water-treatment plants and their many subsystems like special pumps or gas treatment devices.

### 3.2. Existing engineering curricula

A detailed analysis of existing curricula has lead to the following conclusions:

- The nominal *overall duration* of engineering studies at the diploma level is in a range of *4 ... 5 years*.
- All curricula require completion of a *diploma thesis*; its duration ranges from a *minimum of 3 months to a maximum of 6 months*.
- Most programmes require students to spend some time in industry; there are considerable differences, however, in the duration of such *practical trainings*, and how the training periods may be accommodated within the curriculum.
- The standard (and usually only) degree awarded after successful completion of a first phase of studies is a *diploma*. Successful completion means that such a degree is generally recognized on the labor market. So far, *no intermediate degrees* (i.e. at a level equivalent to a Bachelor's degree) are awarded. Through additional attributes, titles can be recognized as belonging to a specific type of institution, e.g. *dipl. Ing. ETH* vs. *dipl. Ing. FH*.
- The role of governmental agencies and recognized professional societies in defining and recognizing such titles is much less pronounced in Switzerland than in other countries, where denotations such as "registered engineer" or "chartered engineer" exist[5].
- There is a clear majority of curricula with a *2+2 structure*, i.e. a *Grundstudium (premier cycle) of 2 years*, followed by a *Vertiefungsstudium (deuxième cycle) of another 2 years* (not counting extended periods of practical training, and often excluding diploma thesis work).

---

[5] In fact, a *Register der Ingenieure* exists in Switzerland as well; its relevance for the job market, however, is rather low.

## 4. A PRELIMINARY COMPARISON OF EXISTING EE CURRICULA AT LEADING EUROPEAN INSTITUTIONS

In the framework of the so-called IDEA cooperation between four leading technical universities, a thorough comparison of electrical engineering curricula was made [5]. This comparison was based on the usual information such as lists of courses showing titles and allocation of time, syllabi, detailed description of course contents and goals, etc. At first sight, there appeared to be considerable differences in the curricula involved, with e.g. physics playing a major role in one curriculum and being virtually absent in another. However, when extending the comparison to a close look at the level of examination tasks, it soon became clear that both *content and quality* at the four institutions are *strikingly similar* – if the *level attained at the end of the third year* is looked at. Obviously, the differences observed in a first round of comparison were mainly due to circumstances like (1) different interfaces to secondary education; (2) different approaches in presenting the underlying sciences (e.g. physics, mathematics, computer science, materials, etc.), i.e. either in a segregated way (e.g. taught as "service" courses by other faculty) or in an integrated way; and (3) different sequencing of subjects.

Despite all these differences, there appears to be *reasonable convergence of curricula towards the end of the third year*, as illustrated in figure 2.

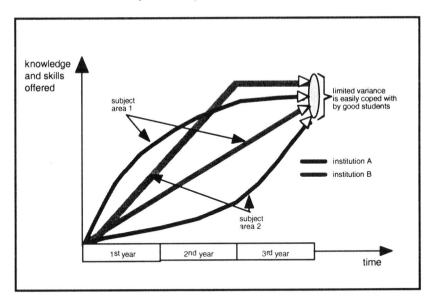

*Figure 2. Convergence of engineering curricula in terms of skills and knowledge level when different institutions are compared.*

Since most curricula offer a limited number of specializations within the following 2$^{nd}$ cycle, divergence is increasing again after the 3$^{rd}$ year even within one institution.

Thus, in the specific case of Electrical Engineering, convincing evidence was found that the *Bologna* proposal with a three-year first cycle is indeed a reasonable approach to curriculum structuring. In fact, such comparisons have been made for other disciplines as well, with similar outcome [6]. Moreover, experience with mobility programmes like *Erasmus* provides additional support for this hypothesis.

Finally, a very important argument was raised in discussions with representatives from various engineering faculties: The spread in capabilities observed for a large group of students is much wider than the differences in curricula at different institutions of the same type and ranking. Thus, figure 2 implies also that *we should not be overly scrupulous when harmonizing curricula* with mobility as our motive for standardization.

## 5. THE PROFILE OF UNIVERSITY LEVEL ENGINEERS

*5.1. General observations*

With a new federal law [7], Switzerland has established in 1995 a framework to convert most of the previous "Technikum" type institutions into what is called in German *Fachhochschulen*. In the engineering field, we are concerned with some 20 so-called *Höhere Technische Lehranstalten (HTL) (École Technique Supérieure, ETS)* or *Ingenieurschulen (École d'Ingénieurs)*, which now form part of one of the presently 7 *Fachhochschulen*. Thus, a typical *Fachhochschule* comprises usually several of the previous HTL along with previous HTL type institutions in other fields such as business or arts. In the message to Swiss Parliament of November 25, 1998 [8], the existing universities and the two ETH's have been regarded together with the new Fachhochschulen as forming a *Hochschulsystem*, with several *Hochschulnetze* (networks covering all three types of Hochschulen, i.e. cantonal universities, the two ETHs and Fachhochschulen) defined in an overlay fashion [9]. Unfortunately, the two documents quoted so far (i.e. [7] and [8]) as well as many other publications (such as [9]) leave it almost completely open whether the three types of *Hochschulen* have discernibly different missions and profiles. In fact, an often quoted "Leitmotiv" is that *Fachhochschulen* are "gleichwertig, aber andersartig" (equivalent, but of different kind), which is interpreted in [8] as meaning that the *educational mission* (Bildungsauftrag) of *Fachhochschulen* is equivalent to that of the universities, whereas the overall *context* is different in the sense that *Fachhochschulen* should aim at bringing science and practical application in close connection. In [10], however, the author concludes that this differentiation will hardly be successful in practice, and in a somewhat pessimistic note, he predicts increasingly blurred boundaries and increasing competition between *Fachhochschulen* and *Universities* (including ETHs).

We propose, however, a well-defined *distinction* between the *profiles* of a *Fachhochschule* type of engineer and one with a *university level education*.

As examples of other countries have shown, regarding the two types of institutions as largely equivalent may be rather *detrimental to the former HTL schools and their graduates, and a blow to the excellent idea to have a two-tier tertiary system* with each pillar having its distinct profile. Specifically, it has been reported from England[6] that graduates from the former polytechnics find it increasingly difficult to get jobs since their educational background no longer matches industry's expectations.

The above findings, however, should by no means be interpreted such that gifted graduates from a FH should not be allowed to continue studies at a university level institution, provided that such a transfer is taking place at a clearly defined interface (see 7.3). On the other hand, the idea of attracting *Gymnasium (gymnase)* graduates to pursue studies at a FH undermines the two-tier tertiary education system.

## 5.2. Professional roles and capabilities

Based on the discussion in chapter 2, we may say that university level engineering programmes aim at educating their graduates such that they are able to

- Assume *leading roles* and contribute significantly to *innovation*; thus, they should usually have a *long-term view*.
- See their work not only in a technical framework, but approach their assignments in a *multidisciplinary* context, with due consideration of human, social, environmental and economical aspects.
- Pursue a career as generalist (i.e. go into *management*) or specialist (i.e. assume an important *staff* function), or become an *entrepreneur*.
- To *reorient* themselves *permanently* with a view to the continuous emergence of new technologies and new approaches to design, development and production.

Essentially, three types of engineers could than be discerned:
    (a)    System Engineers
        aware of the *theoretical* (mathematical and physical) *basis* of most engineering methods and tools; thus, this type of engineer is playing a leading role in *improving methods and tools* (whereas Fachhochschule graduates rather apply these).
        As explained in section 3.5, this type of professional is also needed to approach complex *multidisciplinary* problem areas (as e.g. found when technical processes have a long-term negative impact on the environment), aiming at devising appropriate *strategies* for long-term improvement.
    (b)    Technology Specialists
        perceive the relevance of newly emerging results in the natural sciences (physics, biology, ...) and new *"immaterial"* technologies

---

[6] Quotation from Imperial College's IDEA [5] coordinator: "What the former Polytechnics did well for industry in the past, they no longer do, and what they pretend to do now, they do badly"

(based on mathematics) with a view to develop new types of devices and platforms.
(c) Managers of Large Projects
accept the challenge of *managing complex projects* requiring *unorthodox engineering approaches* (e.g. drawing on results produced by system engineers and technology specialists).

In fact, the common denominator for all three types of roles is that one could say that **University level engineers push the limits of their field[7]**! This has in fact already been visualized in figure 1 with the "push foundations" attribute.

*5.3. Implications for the engineering curriculum*

Given the university level engineering profile sketched above, it is evident that a 3-year first cycle is not sufficient to reach the stated educational goals. Rather, the first cycle is necessary to lay the necessary scientific base, which, on the other hand, can generally not yet be regarded as a professional qualification.

*Thus, a well-rounded university level engineering education calls for a 4-5 year curriculum.* We purposely propose to foresee a *range* of 4-5 years as the actual duration of studies should depend both on the particular aims of a specific programme and possibly also on the capabilities of the individual student when a credit system is used (see also 6.2). Beside the emphasis on a solid scientific basis, due regard should be given to the requirements for non-technical skills as discussed in 2.2; a visible position should be given to the so-called soft skills in the curriculum.

Nevertheless, as stated in the introduction, we see considerable value in defining an intermediate level concluding a first cycle of university studies, i.e. the *Bachelor level* implied by the Bologna proposals. As shown in chapter 4, due to the convergence of the curricula of most leading technical universities at the end of the 3$^{rd}$ year, this level lends itself quite naturally as *a pivot for mobility*, where students go through their undergraduate studies at an institution A, getting a Bachelor certificate by A, transfer to institution B for their graduate studies and conclude these studies with a Master's degree from B. There are several motives for such a so-called *vertical mobility*:

(1) Studying in two different environments undoubtedly increases awareness for the manifold cultural and political environments, thus contributing significantly to forming well-rounded personalities familiar with working and living in an international, multi-cultural context. This is regarded as a very valuable asset for a modern engineer, since many professional tasks have nowadays a distinctly transnational flavour.

(2) There certainly exist many cases where some university excels in a certain specialization, whereas another institution provides top-level education and

---

[7] It was recently observed by the Vice-Dean of the *École Nationale Supérieure de Télécommunications (ENST)* that the failure rate in France for start-up companies in the high-tech field was significantly lower for companies founded by PhD's.

research in a different field. It would be quite natural that these two institutions cooperate, opening their Master and PhD programmes for each other's graduates.

(3) In certain areas (e.g. in informatics and networking), there is an acute shortage of professionals all over Europe. Since the Bologna proposals resemble rather closely the Anglo-American curriculum structure found in many parts of the world, we may tap a source of gifted and ambitious students if the second study cycle is opened for overseas students. As a consequence, at least part of the lectures in the second cycle will be given in English.

Evidently, if this approach is successful, it will confront (in a positive sense) our own students with increased cultural diversity as well.

## 6. A FRAMEWORK FOR A GENERAL UNIVERSITY LEVEL ENGINEERING CURRICULUM

### 6.1. Basic structure and interfaces

Based on the considerations in previous chapters, we propose a curriculum model as illustrated in figure 3.

The essential features of this structure are:

(a) The normal university level engineering curriculum is a 4-5 year integral course leading directly to a Master's degree.

(b) We propose a standard two-cycle curriculum structure, with the first cycle (i.e. the undergraduate part) of three years leading to a Bachelor's certificate, followed by a second cycle (i.e. the graduate part) leading to a Master's diploma. The actual duration of the second cycle may vary between 1 and 2 years, depending on the field and how individual students fulfill credit requirements.

(c) In contrast to the Bologna Declaration, the Bachelor's degree is neither seen as a professional qualification, nor is it a compulsory prerequisite for graduate admission if studies are continued within the same institution.

(d) Students following the integral Master curriculum may simultaneously obtain a Bachelor' degree if specific requirements set up for this degree were met.

(e) The Bachelor's level is primarily interpreted as an intermediate mobility pivot, i.e. an interface where the conditions for a student transfer between cooperating institutions are well defined, or where industrial practice is sandwiched between the two cycles.

(f) Admission criteria for students entering a graduate programme at the Bachelor's level are primarily set up by individual institutions (departments, faculties), i.e. there should not be a comprehensive standardization of knowledge and skill levels across disciplines.

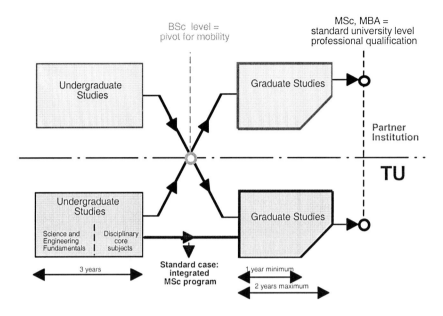

*Figure 3. Basic curriculum structure, showing BSc level with its role as a mobility pivot.*

### 6.2. Duration of studies, credit system

As stated in section 6.1, *use of a credit system is assumed*. There are essentially two motives for such a system:
(a) With a view to mobility, students must be able to provide *proof of their achievements*.
(b) In order to allow very gifted students to follow a *fast track* through the curriculum, a transparent system to evaluate their achievements is needed.

While point (a) is valid for both the undergraduate and the graduate part of the curriculum, a *fast track* option does hardly make sense for the first cycle with its much firmer course structures and interdependencies between courses. Thus, we see a *fixed duration of the first cycle of 3 years*. On the other hand, the characteristics of the second cycle (many course options and specializations, project work, industry practice, etc.) are well compatible with the idea of students moving through this cycle at somewhat different speeds. For example, former *Fachhochschule* graduates transferring to a university level institution may be dispensed from part of project work, allowing them to concentrate on formal courses, or some students may allocate project work to course-free periods. Thus, we recommend that the duration of the *second cycle should not be rigidly specified*; we rather propose a *nominal duration of 1 year minimum and 2 years maximum* (not counting extended practice periods in industry, if required).

The curriculum structure finally chosen by the ETHZ EE department[8] is shown in figure 4.

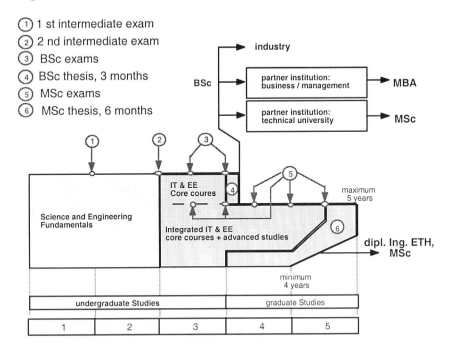

*Figure 4. An overview of the proposed new curriculum at the ETHZ EE department.*

### 6.3. Degrees and examinations

We are of the opinion that the "old" types of curricula leading to *Diploma* or *Lizentiat* degrees should *not* co-exist with the new 2$^{nd}$ cycle leading to a Master's degree. A practical approach will very often consist in retaining the "old" type of degree (e.g. dipl. Ing. ETH, ing. dipl. EPF) as far as the German, French or Italian version of the relevant document is concerned, while - so to speak - the English translation on the back of the same document denotes the level reached as a Master's degree.

In order to differentiate between university level and *Fachhochschule* level degrees, the following denominations could be used:
- *University level*: Bachelor of *Science* (BSc), Master of *Science* (MSc)
- *Fachhochschule*: Bachelor of *Engineering* (BEng), Master of *Engineering* (MEng).

---

[8] Pending a decision of the ETH Council, the ETHZ EE Department will be renamed *Information Technology & Electrical Engineering (IT&EE)*.

The attribute "of Science" has purposely been chosen to point to the scientific basis of engineering studies at university level.

## 7. MOBILITY AND PROFESSIONAL QUALIFICATION

*7.1. Overview*

We see a number of cases for mobility as illustrated in figure 5.

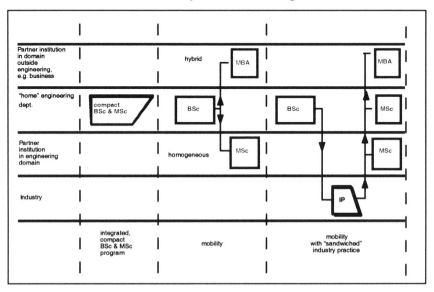

*Figure 5. Different cases for mobility using BSc pivot.*

*7.2. Vertical Mobility within discipline at university level*

This is the preferred form of mobility, responding to our considerations in section 5.3. According to our statement (e) in section 6.1, it must be emphasized that *admission criteria* for students entering a graduate programme at the Bachelor's level are primarily *set up by individual institutions* (departments, faculties), i.e. there should *not* be a comprehensive standardization of knowledge and skill levels across disciplines. Also, there should be some flexibility in defining the BSc pivot depending on the target Master cycle. We will come back to this point in section 7.4.

## 7.3. Transfer to and from Fachhochschulen

There exist many cases where graduates **from** *Fachhochschulen* did very well when continuing studies at university level. In many such cases, these people become aware at *Fachhochschule* that their background in theory and in natural sciences is just not sufficient to allow them to become what we have called in section 5.2 "engineers able to push the limits of their field"; thus, they are highly motivated to acquire such a background through studies at university level. Presently, the prevailing solution for a *transfer from Fachhochschule to ETH/university* exhibits the following elements:

- *Fachhochschule* graduates are accepted if their grades obtained at graduation are above a certain limit.
- Such students enter at the start of the $3^{rd}$ year, i.e. after *Grundstudium*.
- In order to be accepted, applicants have to demonstrate (e.g. through an exam equivalent to the $2^{nd}$ Vordiplom at ETHZ) that they master the subjects of the university level *Grundstudium*.

Such a scheme must be maintained even if graduates from *Fachhochschulen* get a *Bachelor's* degree. There might be exceptional cases where the present rules could be relaxed in the sense that Fachhochschule Bachelors could be accepted to directly enter a university Master's program, provided that they make up for missing parts of the university Grundstudium through special efforts (accepted only if exams are passed).

So far, transfers in the opposite direction, i.e. from universities **to** Fachhochschulen, have been rather rare; primarily, they were made by students failing in their intermediate exams (first or second *Vordiplom*). However, in those cases where more emphasis is given to the scientific and theoretical basis of university level engineering education in the future, such transfers might become more frequent, and consequently, well defined transfer mechanisms will have to be defined.

## 7.4. Hybrid curricula at university level

Fields like *management* and *finance* look at their subjects more and more with a *system view*, with mathematics and notions of the natural sciences becoming increasingly important if appropriate models are sought[9]. Thus, graduates from an undergraduate engineering program should in principle be well prepared to enter a graduate level program in these fields, provided they make up for some deficiencies of their technically oriented *Grundstudium* when compared to the undergraduate program of e.g. a business school. There is an elegant *proactive approach* to solve this problem, provided that the undergraduate engineering program offers some options for the students, e.g. in the way she/he fulfils the requirements described in

---

[9] As an example, we may consider the approaches now used to model processes on the stock market (e.g. using elements from the theory of fractals). Many such models are very familiar to networking engineers, where e.g. the self-similar properties of data traffic have been identified a few years ago.

section 6.3 about non-technical skills. Such an approach is illustrated in figure 6, where it is assumed that introductory courses in management, finance etc. are offered in the 3$^{rd}$ year.

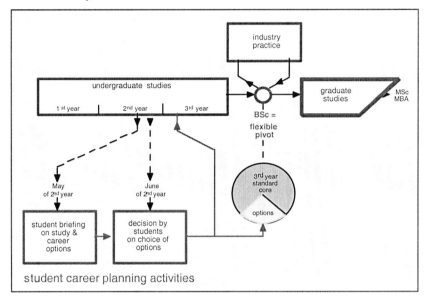

*Figure 6. Defining the BSc level in a flexible way through 3$^{rd}$ year options.*

Thus, students deciding to follow a *hybrid curriculum* as shown in figure 5 would compile their 3$^{rd}$ year program in such a way as to optimally interface to the target institution for their graduate program. Of course, this approach is also feasible to overcome any divergencies encountered with vertical mobility when students stay within their technical field of choice.

We believe that graduates from such hybrid BSc/MBA programmes would fare very well in tasks such as engineering management, technology management or risk assessment (e.g. for insurance companies), since their view of the scientific basis of technical systems and their systematic approach in analysing complex situations was thoroughly schooled in their undergraduate engineering courses.

## 7.5. Curricula with extended industrial practice

It has often been argued that, despite our efforts to explicitly define the undergraduate part of the engineering curriculum as a preparatory phase for graduate studies (i.e. that the BSc level would not constitute a professional qualification), industry would try to lure students away from university at BSc level, above all in times of acute shortages on the labour market (as it is the case for many high-tech professionals today). We think that this argument can be countered as follows:
• Students leave university already today during their undergraduate studies for the reasons given above. However, they do this without any certificate, and

moreover, managers of high-tech enterprises consistently state that such collaborators may well serve as stop-gap measures for specific tasks, with however no long-term perspective in the firm. Thus, we prefer to have such students leaving with a Bachelor's degree, allowing them to resume graduate studies after a 1- or 2-year practice in industry.
- We assume that there will be increasing political pressure such that students themselves should accept a bigger share of the cost of education, e.g. through larger contributions by their parents, through scholarships or with the aid of repayable grants. The model proposed in figure 5 shows that the idea of the Bachelor level mobility pivot could serve very well to partially solve this problem, since students could "sandwich" e.g. a year of paid industrial practice in between the undergraduate and the graduate part of their studies. This might have the additional benefit that such students usually return well motivated to get the education they found missing when the problems encountered in industry ought to be approached systematically.

## 8. CONCLUSIONS

The EE department at ETH Zürich will open its new curriculum in the next Winter semester[10], with students presently in their 1$^{st}$ and 3$^{rd}$ semester profiting from the new possibilities through some transition rules. Several other departments at ETH Zürich seriously consider the introduction of new curricula, many more or less influenced by the EE department's model. Both the detailed study plan and the exam regulations are available in electronic form [15].

In the meantime, the approach taken by the ETHZ EE department was taken over as a model by a working group set up by the *Conference of the Rectors of Swiss Universities (CRUS)* [2].

## 9. REFERENCES

[1] *The Bologna Declaration: The European Higher Education Area.* Joint declaration of the European Ministers of Education, convened in Bologna on the 19$^{th}$ of June 1999. See e.g.
http://www.unige.ch/cre/activities/Bologna%20Forum/Bologne1999/bologna%20declaration.htm
[2] Rektorenkonferenz der Schweizer Universitäten: See e.g.
http://www.crus.ch/deutsch/shrk/Projekt/Bologna.html
[3] CRUS Generalsekretariat, Bologna-Koordination: *Einsetzung und Betreuung fachbezogener Arbeitsgruppen im Zusammenhang mit dem Bologna-Prozess in der Schweiz* (Dokument 00-005).
[4] Waite Maurice (ed.): *The Oxford Colour Dictionary.* Oxford University Press, 1995.
[5] IDEA League: *Report on Comparison of Curricula in Electrical Engineering.* IDEA League secretariat, Delft, 30-10-00.
[6] Oral communication by Prof. Peter Marti (ETHZ, Baustatik und Konstruktion).
[7] Bundesgesetz über die Fachhochschulen vom 6. Oktober 1995 (Fachhochschulgesetz; FHSG). SR 414.71.
[8] Botschaft über die Förderung von Bildung, Forschung und Technologie in den Jahren 2000-2003 vom 25. November 1998. Dokument 98.070.

---

[10] As the introduction of the new degrees requires the consent of the ETH Council, formal approval of the new study plan should take place in late May 2001.

[9] Kleiber Charles: *Die Universität von morgen. Visionen, Fakten, Einschätzungen.* Staatssekretariat für Wissenschaft und Forschung, Bern, 1998.
[10] Dubs Rolf: *Der Aufbau von Fachhochschulen in der Schweiz: Stand und Probleme.* VSH-Bulletin Nr. 2/3, August 1998.
[11] Weinert Donald G.: *The Definition of Engineering and of Engineers in Historical Context.* This paper can be downloaded from http://books.nap.edu/books/0309036399/html/71.html#pagetop.
[12] IDEA League: *Report on Virtual BSc in Chemistry / Chemical Engineering.* IDEA League secretariat, Delft, 30-10-00.
[13] For ETHZ, see e.g. http://www.mobilitaet.ethz.ch/de/outg/outv/programme.html, for EPFL, see http://www.epfl.ch/RI/Reseaux.html.
[14] ABET: http://www.abet.org/downloads/2001-02_Engineering_Criteria.pdf
[15] A link to these pdf-Files is given on: http://www.ee.ethz.ch/teaching/.

*Albert Kündig*
*ETH Zürich*
*EE Department*
*Zürich, Switzerland*

S. S. MELSHEIMER

# AN AMERICAN OPINION OF THE EUROPEAN ADAPTATION OF THE B.S./M.S. DEGREES

**Abstract.** A major impediment to articulation of U.S. and European degree programs has been the incompatibility of the European Diplom/Doctor scheme and the U.S. BS/MS/PhD scheme. The introduction of the bachelors/masters structure in European institutions promises to open the opportunity for U.S. B.S. graduates to travel to Europe for masters level studies, and it also suggests the possibility to develop true international engineering programs involving close articulation and possibly dual degrees. The growing introduction of English language instruction (primarily at the master's level) in Europe also removes a significant barrier for US engineering students. However, there remain real differences in the functioning and format of the European BS/MS degrees versus the BS/MS structure in the US.

## 1. INTRODUCTION

With the growing integration of the global economy, there is a real need for globally educated engineers. One desirable form for this to occur is by having interchange of students between the U.S. and Europe. However, the structural difference between the U.S. BS/MS/PhD system and European Diplom/Doctor system has hampered mobility of students between the two systems. Part of the problem has been poor understanding of the opposite system, especially by US faculty. However, there are real problems in articulation. For example, a U.S. student with a B.S. is generally not qualified to enter a European doctoral program, and there has been little opportunity to gain the incremental education (perhaps 1-2 years) needed to achieve equivalency with the Diplom other than by obtaining an MS in the US. At that point, US students desiring a doctorate are likely to pursue a PhD within the US. In addition, U.S. students are often lacking in language skills. European students possessing a university Diplom are well prepared to pursue a US doctorate, but many US universities do not recognize the Diplom as equivalent to a US masters. Thus, they may require entering a masters program prior to admission to the PhD.

The initiative of the European institutions to move to a BS/MS system that is at least similar to the US system promises to reduce or eliminate these problems, and is welcomed in the US. However, there remain some real differences that may still limit the goal of increasing mobility across the Atlantic, especially from the US to Europe. This paper provides an American perspective on the proposed programs, and thoughts on the problems that may continue to inhibit Transatlantic articulation. First, a review of some salient characteristics of US Engineering Education, and a view from the opposite side of the Atlantic of the "traditional" European structure.

## 2. U.S. ENGINEERING EDUCATION

U.S. engineering schools are usually housing within a comprehensive university. While these schools have a great deal of autonomy in defining degree programs, various accreditation standards impose a fairly uniform program structure. In addition, institutional policies provide further constraints on the degree programs.

### 2.1. Institutional Accreditation

Each university or college in the U.S. is accredited by a regional agency (for example, the Southern Association of Colleges and Schools (SACS) accredits schools from Virginia to Texas). Each regional accreditation agency is independent. While they accredit entire institutions rather than individual degree programs, they do impose general standards for degree programs.

- Bachelors programs require four years of study, typically organized as 8 semesters of 15-18 credits per semester (120-140 credits for a B.S.). This includes a general education core of about one year of study.
- Masters programs require a minimum of one year of study beyond the bachelors or equivalent, and may be research oriented (typically 24 credits plus thesis) or coursework oriented (30-36 credits).
- Doctoral programs are defined in terms of achieving a research objective rather than a time period of study, but typically require 3+ years beyond the masters and usually require some additional coursework beyond the masters.

### 2.2. Program accreditation

ABET, the Accreditation Board for Engineering and Technology, sets standards for degree programs providing qualifications for engineering practice (professional licensing also requires examinations and a period of practical training). Institutional accreditation is a prerequisite for professional program accreditation.

- ABET accredits the "first professional degree" as defined by the engineering school. This can be either the BS or MS, but with rare exceptions it is the BS.
- ABET provides general specifications on math and science content, engineering content, a general education component, plus disciplinary specifications on topics that must be addressed within the curriculum.
- Internships are optional (some engineering colleges have mandatory co-op programs, others have voluntary internship and/or co-op programs) and not part of the "credits" for the degree.
- All programs must include a major engineering design project, but an independent thesis is not typically required (except for Honors degrees).

With the new outcomes-oriented ABET EC 2000 Accreditation Criteria (Table 1 summarizes key criteria related to the present discussion), each program is required to define detailed educational objectives, a curriculum to achieve them, and a system to demonstrate achievement of these objectives. It is required that program constituents be engaged in developing and evaluating the educational objectives.

This provides great flexibility in accreditation of programs with considerably different goals. For example, programs may be designed to be more practice-oriented (e.g., if most students terminate studies at the B.S.) or theoretical (if it is expected that a larger fraction continue to graduate study).

*Table 1. ABET EC 2000 Criteria (Selected BS Level Criteria)*

Criteria 2 - Program Educational Objectives. The program must have detailed educational objectives, curriculum designed to achieve them, system of evaluation that demonstrates achievement of objectives.

Criteria 3 - Program Outcomes. Graduates have demonstrated ability to
   (a) Apply knowledge of math, science, engineering
   (b) Design & conduct experiments, analyze & interpret data
   (c) Design a system, component or process to meet desired needs
   (d) Function on multi-disciplinary teams
   (e) Identify formulate and solve engineering problems
   (f) Understand professional and ethical responsibility
   (g) Communicate effectively
   (h) Broad education so can understand impact of engineering solutions in global/societal context
   (i) Recognize need for and able to engage in life-long learning
   (j) Knowledge of contemporary issues
   (k) Use techniques, skills and modern tools of engineering practice

Criteria 5 - Curriculum Content (4 years total) must include
   1. 1 year math and science appropriate to the discipline
   2. 1.5 years engineering topics appropriate to the discipline culminating in a major design experience
   3. ~ 1 year general education consistent with program objectives
   4. ~0.5 year unspecified (used to meet specific program or institutional objectives)

Criteria 8 - Program Criteria. Specifications on curricular topics and faculty qualifications are provided for each recognized discipline.

*2.3. Institutional Constraints*

Typically, institutions have broad "general education" programs and other policies that are imposed on all degree programs, especially at the BS level. Some such policies also exist at the graduate level. Typical BS level policies include
- Minimum course credit requirements in English composition, humanities, social sciences, mathematics, etc.
- Minimum and maximum numbers of credits allowed for a degree.

In addition, state supported institutions may have curricular policies imposed by the state legislature or a state governing board. The consequence of all these rules is that transfer of a student from a dissimilar educational system into a U.S. B.S. degree program (or, establishment of a dual-degree program) may be difficult. However, the situation is generally more flexible at the graduate (MS or PhD) level since there are fewer accreditation and general education rules.

*2.4. Fee Structure*

At the BS level, students in American schools typically pay a moderate to high tuition charge (from about $3K to over $20K per year). Scholarship programs may cover some or all of this for very talented students and for financially needy students.

Well-qualified students in research-oriented masters programs often receive assistantships (or fellowships) covering not only tuition, but also stipends large enough to cover living expenses. Students in coursework masters programs may pay substantial tuition, though assistantships or fellowships are sometimes available in technical fields. Note that in non-technical masters (e.g., the M.B.A.) it is not normal for students to receive assistantship/fellowship support.

At the PhD level, it is the norm that engineering students receive support through assistantships or fellowships that cover living costs as well as tuition.

## 3. TRADITIONAL EUROPEAN SYSTEM (DIPLOM/DOCTOR)

These degree programs are governed by national standards, and thus there are variations between countries. However, to an outside observer it appears that the key characteristics are general even though details may differ.

*3.1. Degree Characteristics*

The Diplom is the first professional degree, and provides qualifications for professional practice. A certificate (the Vordiplom in Germany) is often awarded after completion of basic level studies at the end of the second year. However, in many countries different Diplom degrees exist in two different types of institutions. One is the university, which has research as a major mission and which grants doctoral degrees as well as the Diplom. The second type are institutions such as the

German Fachhochschule, institutions that offers only the Diplom and has a more limited research emphasis. I will refer to these as technical colleges.

- The university Diplom has a 5 year nominal duration (including ~ 1 year for internship and thesis) and approximates a U.S. masters (although it is not always accepted as such in U.S. because it is a "first degree"). The university Diplom typically has a relatively fundamental (theoretical) focus.
- The technical college Diplom has a 4 year nominal duration (including ~ 1 year for internship and thesis). It approximates U.S. B.S. in duration and scope of technical content, and typically has a more practical orientation than a university Diplom program.
- The Doctor (university only) is very similar in scope and goals to the U.S. Ph.D. A university Diplom or equivalent is usually required for admission.

*3.2. Fees*

At present, tuition fees in Europe (BS/MS as well as Diplom) are nonexistent or negligible.

*3.3. Issues with Diplom/Doctor Structure*

Articulation with the US degree structure is problematic because the U.S. B.S. is not adequate for entry into a European doctoral program and no good "bridge" program has been available in Europe. A U.S. M.S. would typically be acceptable for entry into a European doctoral program, but at that point few U.S. students would make the transition. Further, the university Diplom is not always accepted as an MS equivalent in the U.S.

A second issue is the use of the same degree label ("Diplom") for two significantly different degrees. Not only do they have different educational goals (more theoretical versus more practical), they have different nominal durations of study. Further, while they have some level of legal equivalence, the technical college Diplom is generally not accepted for entrance into university doctoral programs.

Finally, the university Diplom is a rather long and inflexible first degree, with a duration that is often longer than the nominal 5 years. It is a very "strong" degree program, but it may be "overkill" for many engineering positions. It also inhibits student transitions into other professions (e.g., management, medicine, law).

By way of comparison, the 4-year US BS degree is the terminal degree of most US engineers and provides full qualification for professional licensure. Upon completion of the BS, those individuals seeking to enter more highly technical careers may elect to pursue an engineering MS (and sometimes then a PhD). Other graduates may enter an MBA or other management program, while others enter other professions by attending law school, medical school, etc. The latter fields offer significant opportunities to engineers and others with a technical B.S. degree.

## 4. EUROPEAN BS/MS SYSTEM

The new European BS/MS structure is intended to address these and other issues with the Diplom. Unfortunately, one attribute that may be inherited from the Diplom is the existence of "equivalent" (but different) B.S. degree structures within the research universities and the technical colleges. These programs are still in an early stage and implementations do vary somewhat at different institutions, but the current situation appears generally as follows:

- The university BS appears to be standardizing at a nominal 3 years duration. This may include an internship and/or thesis, but is heavily weighted to coursework. It would appear to closely approximate a U.S. BS, given that the first year of a U.S. BS is largely material presumed from secondary school in Europe. It is likely that the university BS will retain the relatively fundamental/theoretical bent of the university Diplom.
- The technical college BS also appears to be standardizing at a 3 year duration, but typically includes about one year devoted to internships and thesis work. Thus, it appears to be a bit short of the U.S. BS in course content. On the other hand, the U.S. BS does not normally include a thesis project. The technical college BS is expected to retain a more practical orientation.
- The MS appears to range from 1-2 years, and to reasonably closely approximate the U.S. MS. The European MS seems to typically have a coursework emphasis. Students typically do not receive stipends (but, tuition is free or nominal). Many MS programs are being offered in English to attract foreign students. Presumably, there may be a more practice-orientation in technical college MS programs compared to the university MS.

Figure 1 compares the overall duration of the various degree schemes, with programs extending through the BS & MS (or, the Diplom). In so doing, one year is added on the European side to reflect the longer duration and higher level of typical students' secondary school education. Some variation exists both in the US and in Europe in the length of an MS program. A nominal value of 1.5 years is used for the US MS, though a research-oriented masters is often longer and a coursework-only masters may be as short as one year. A value of 1.5 years is also used for the European technical college MS. For the university MS, a value of 2 years is used (this provides a BS+MS duration that is the same as the university Diplom duration, which appears to be the intent at most universities).

Most European programs (BS/MS and Diplom) include at least a half-year of internship or practicum. In the U.S., such a practicum is not normally counted as part of the basic degree requirements (total credits), and is optional in most institutions.

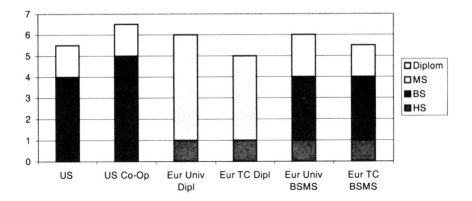

*Figure 1. Comparison of Program Duration*

Programs with mandatory co-op/intern schemes typically have a nominal duration of 5 years, and voluntary co-op programs also require 5 years to complete. Thus, a second US entry is shown which reflects this Co-Op option.

It is relevant to compare the European BS/MS programs against the general ABET curriculum expectations. These are

- ABET basic level (BS)
    - 1 year math, science
    - 1.5 years engineering
    - general education as specified by the institution
    - .5 years of topics chosen to meet program objectives (implicit)
- ABET advanced level (MS)
    - 1 additional year engineering and related topics

Much of the "general education" content and about one-half year of the math and science included in US baccalaureate programs are covered in European secondary schools. Thus, the nominal 3-year BS structures at both technical colleges and universities should cover the scope of math, science and engineering required by basic level accreditation standards (to actually achieve accreditation, of course, additional detailed standards must be met). Similarly, the MS at both types of institutions should provide sufficient additional engineering and related topics to meet advanced level standards. Note that the flexibility inherent in the EC2000 Criteria would enable both types of institutions to meet accreditation requirements.

## 5. ISSUES WITH THE EUROPEAN BS/MS STRUCTURE

Clearly, the European BS/MS structure articulation with the US degree structure is significantly improved. Moreover, the growth of program offerings in English

(especially MS programs) helps overcome the "language problem" in the US. Yet, there remain some issues that may inhibit flow of students between the two systems. Some of these include:
- If university and technical college BS degrees are not equivalent for entry to European MS programs, what will their status be in the U.S.?
- If the technical college MS is not accepted for entry to European doctoral programs, its acceptance for US PhD admission may be questioned.
- Good U.S. students receive stipends (assistantships/ fellowships to cover living costs as well as tuition) in U.S. M.S. programs. Few will be attracted to the European MS without comparable provisions to cover living costs.
- What is the future of the Diplom? Resistance to elimination seems strong!
    - "Hybrid" ideas have appeared, e.g.
        - Students allowed to choose either Diplom or BS track after 2-3 years of study. I.e., programs are the same at least through Vordiplom level.
        - University MS programs advertising admission of
            - University Diplom students after 3 years study (de facto BS)
            - Students with technical college Diplom
            - Students with U.S. B.S.
    - Maintaining both systems seems like the worst of all worlds - but, maybe it's the best!

While the question is raised above as to the acceptance of technical college BS and MS graduates into U.S. M.S. and Ph.D. programs, most U.S. engineering schools are likely to be open to admission of these students. They will look at the academic performance and capabilities of the individual student as well as the curriculum content of the specific B.S. institution. The variability of the orientation and quality of individual U.S. B.S. programs is at least as great as is likely to be seen among the various European BS programs. U.S. schools will evaluate the performance of admitted students, and this experience will guide future admissions. It is not clear whether this perspective will prevail in Europe - but, from an outside perspective it would appear that it would be very healthy if it did.

## 6. CONCLUSIONS AND PERCEPTIONS

1. The European BS/MS already appears to be fostering the increased flexibility that was one of its objectives, as illustrated by university MS programs that seek applicants from U.S. B.S. programs, technical college Diplom programs, and university Diplom students who have completed three years study as well as university B.S. programs.

2. The European BS/MS will be well received in the US, but a single system (i.e., elimination of the Diplom in favor of the BS/MS) is desirable though not essential. Further, standard nominal degree lengths are highly desirable (e.g., BS=3 years, MS=1.5 years).

3. Treating technical college and university BS and MS degrees as equivalent credentials for further study is highly desirable. This does not suggest that the two types of programs should be identical - individual institutions may have quite different program objectives, and students, employers and other constituents can assess program appropriateness for their needs. However, strong students from technical colleges are likely to be successful in pursuing higher degrees at universities, and the "system" should provide this flexibility.

4. A European-wide flexible accreditation system (similar to EC 2000) is desirable to help ensure compatibility between educational systems in various countries. However, the degree level for professional certification/licensure is a separate issue - this could be the B.S. as in the U.S., or it could be at the MS level if assessment of qualifications needed for professional practice so indicates.

5. It would be highly desirable to have maximum compatibility between a European accreditation system and the ABET Criteria (and, accreditation systems being developed in other locations) to facilitate integrated programs that potentially offer the ideal educational environment for the "global engineer".

6. U.S.-European program articulation opportunities do exist. These include
   - Student exchange and/or collaborative projects providing certificates to augment the basic B.S. (or M.S.) degree awarded in either the US or Europe.
   - BS-MS articulation (collaborative programs with the first degree in either the U.S. or Europe and the second degree on the other continent).

7. However, B.S. level dual-degree programs face formidable obstacles, such as
   - Institutional curricular requirements for BS degrees that are often inflexible.
   - Accreditation systems in the U.S. and Europe may impose conflicting requirements.

Finally, a prediction: while the traditional Diplom may not disappear quickly, I believe the flexibility inherent in the BS/MS structure make it more attractive and will result in the ultimate demise of the Diplom. This flexibility will prove advantageous to both students and employers, allowing depth and breadth of education to be better tailored to the needs of both.

*Stephen S. Melsheimer*
*Clemson University*
*Clemson, SC, USA*

# PART V.

# DEVELOPING PERSONAL SKILLS TO BE A GLOBAL ENGINEER

B. RICHERZHAGEN


# THE YOUNG ENTREPRENEUR'S EXPERIENCE OR: «WHAT CAN THE UNIVERSITIES DO TO MAKE OF A STUDENT A SUCCESSFUL GLOBAL ENGINEER?»

## 1. INTRODUCTION

First, I want to try to explain what is a Successful Global Engineer under the vision of a young entrepreneur so that the question should be «What is a *successful global engineer-entrepreneur*?». Then, I want to tell about my personal *motivation* to become an entrepreneur, which is to *create*, to build up something.

Some skills you should have from birth: *modesty, optimism, patience, perseverance, creativity and the faith in yourself.* Others have to be learned during childhood and youth: *attentiveness, politeness, self-assertion, determination, efficiency, social group capability, and languages.*

The following skills have to be taught at the university: (Recommendations are given what should be changed in order to educate more successful global engineers.)
*Technical knowledge.* Redefine the curriculum at the technical universities, shorten the study length, and improve the education capabilities among the university teachers.
*Knowledge of foreign cultures.* An international study would open the horizons of the engineer which is an important benefit.
*Practical experience in manual capabilities.* A short practice in a workshop before the start of the study would be very useful to avoid having two left hands.
*Professional experience.* Regular visits of manufacturing companies would be an enormous enrichment of the study. This aspect cannot be valued highly enough.
*Health.* The importance of the skill to maintain good health is underestimated. Today's 40-years old managers show illnesses as 60-years old counterparts in the past.
*Economical knowledge.* Not learning by heart of definitions but economical basic knowledge from the practice allowing the engineer to find his way in the jungle of the economic world.

## 2. WHAT IS A «SUCCESSFUL GLOBAL ENGINEER»?

There are a lot of aspects, which should be considered. I would like to analyze the Successful Global Engineer as seen from the angle of a young entrepreneur, therefore, I reword the question: What is a successful global engineer-entrepreneur? The answer is simple: An engineer who creates a valuable commercial activity. As simple is the answer; so complex is the way to become a successful entrepreneur. The way is (from my experience) long, full of obstacles and setbacks, sapping one's energy, and sometimes quite desperate. But at the end of this path, there is a satisfaction hard to match. The creation of a company is one of the highest challenges, which an engineer can accept as a personal goal.

## 3. WHAT ARE THE MOTIVES BEHIND THE DECISION TO BECOME AN ENTREPRENEUR?

I want to give you a very personal response: besides the foundation of a family, the most important purpose in my life: to CREATE, to CHANGE something in the world, so that some traces will remain even when I belong no more to the living. This gives me the impetus to accept all the «tortures", which are inevitable with this decision.

## 4. WHAT ARE THE SKILLS TO BECOME A SUCCESSFUL ENTREPRENEUR BEFORE STUDIES?

### 4.1. Skills from Birth

First, there are some skills with which you have to be endowed from birth, and which you cannot learn later.

*Modesty*, because during the first years as entrepreneur, you cannot expect that you will have a salary such as your colleague who has started his career at Daimler-Benz. To the contrary, you will be poor as a church mouse because you have to spend all your money on your company.

*Optimism*, because even in the most difficult situation, you have to believe that you will solve the problems. «It will become better» must be your most repeated sentence said to yourself.

*Patience*, because you will not found a 1Billion$-company in 3 or 5 years. It is a long course, and in any case, it takes each time longer than you would have thought. You can be happy if your company has survived the first years.

*Perseverance* and *tenacity*, because some customers can only be obtained when you solicit them during a long time. Or in negotiations with your suppliers. «Never give up» should be the second most repeated sentence.

*Creativity* is developed from the pleasure in playing. The concentration of a child on a toy is the same creativity you need when you are an entrepreneur.

*Faith in your self.* But the most important skill is the deeply ingrained *faith* that you will succeed in the end.

*4.2. Skills learned in childhood and youth*

Other skills must be learned in childhood and youth at school:

*Attentiveness*, because you have to understand 100% what your customer needs. It happens so quickly that you offer a product or a solution to your customer, which does not correspond to his demands. He will never buy from you. The same attentiveness has to be directed towards your employees. You must investigate in an early stage internal difficulties and misunderstandings.

*Politeness*, because even in the century of electronic communication, business is still concluded between human beings. The decision to buy at one company and not at another is sometimes irrational. One person has to decide to buy or not to buy, and it is not the result of a mathematic calculation, but the result of a series of impressions leading the buyer to make his decision.

*Self-assertion* towards your collaborators because you are the head of your company. You have to defend your ideas against internal resistance, or towards your suppliers because they do not deliver in time, or towards your customers in sales negotiations.

*Determination*, because you can do a thousand things, you can work 48 hours a day, but you have to concentrate on the most important things which are the following questions: what can I do today to satisfy my customers or how to win over new customers. Never forget to aim at what is before your eyes.

*Efficiency*, because you have so much to do that you have to be very effective in your daily work in order to accomplish a maximum.

*Social group capabilities* like helpfulness learned in youth organizations are good for the capability to cooperate with others. The willingness of responsibility is a basic condition to become an entrepreneur.

*Foreign languages*, because languages are a basic tool today which greatly simplify understanding. I expect my employees to speak, read and write three languages:

English, German and French. Even at the end of the 21$^{st}$ century, we will not have a common language. Who has the job to teach languages? Schools. Unfortunately, the way in which our children today learn languages is completely inefficient. I studied English 8 years at school. After a half- year of private instruction within a French family two evenings per week, I spoke French better than English. Language instruction has to be reformed completely.

## 5. WHAT ARE THE SKILLS, WHICH HAVE TO BE LEARNED AT THE UNIVERSITY?

*5.1. First of all: technical knowledge and methods*

Not too specific, rather a «General Study». A good engineer does not know everything from memory, but he knows, where and how he can find the response to his question or the solution to his problem. The students should learn much more methods at the University rather than definitions. When I learned definitions for an exam, the day after the exam I had forgotten them. But methods I have applied several times I have retained in my mind.

Today, I realize that I use only a fraction of the technical knowledge I learned at the University. Isn't this a terrible revelation? Do Universities teach the wrong knowledge? Or should be the academic length reduced to a couple of months?

I think not, because for each engineer this fraction of the total learned knowledge is different. The University cannot and shall not diversify so much that I can choose in the beginning for example the orientation «How to optimize the movement of the windscreen washer?» because I'll get upon finishing my study, a job at BMW where I have to work on this technical detail. In any case, I would lose my job, if my boss asks me to change my activity to analyzing the electric windows.

That means, it is normal that we do not use almost all the University knowledge in a specific job. The industry needs generalists not specialists. The special knowledge will be learned during the job. For me, a perfect employee has good basic knowledge in Physics, Material science, Mechanics, Electronic and Information Systems.

*5.2. Curriculum*

So, everything is OK? No, because I have the impression that the choice of the contents of the academic education is not OK. The University forms engineers for industry. Only a minority of engineers decides to continue an academic career. Therefore, the industry has to participate in the decision which knowledge has to be teach.

What should be made better and how regarding the technical knowledge? My recommendation: analyze the real needs in industries among today's engineers and make a resume of which knowledge and other important skills, which I mention later, they really need.

This resume will show which knowledge is strongly demanded and applied in industry and what other knowledge is much less important. Industry and University have to sit at the same table and discuss a revision of the University curriculum.

The curriculum has to be in a continuous change, in the same way as the industry changes. I have the impression that the University decides too much without consulting industry, which has to work with these engineers. Too often, Universities teach passing over industry's requirements.

Let's have a look at the US-Universities: thanks to the market driven education system, they have to adapt their curriculum to the market, that means the industry. There is healthy competition between the Universities. In consequence of this, they are in constant change in direction of improvement to the advantages of the students.

The interest of foreign students to attend a European University has greatly diminished. It is becoming more and more urgent to change radically the whole structure of the Universities – not to Americanize them, because the US-model is not at all perfect (which is too un-social: the industry may lose valuable engineers because their parents are not in the necessary financial position).

## 5.3. Length of Study

The German length of study is one of the longest in the world. That is an unacceptable situation. I still remember: here at Aachen, the average length of study of Maschinenbau (machine construction) was 14 semesters or 7 years. Several students needed even 8 or 9 years. Are the German students lazy? No, certainly not. It is a consequence of a completely inappropriate education system, which is too inflexible. Once you have missed or failed one exam, you are immediately out of sync with the prescribed academic schedule and you will lose immediately one year.

A student having studied 7 years is not better than a student having terminated after 4 years. After 4 years and 6 months of diploma work, an engineer has to complete university, if he was an excellent student or a student having difficulties with exams. The student has to get the chance to repeat a failed exam after a couple of weeks, not after one year.

Such a change would not at all influence the quality of the education.

## 5.4. Disorientation of today's university entrants

When I have made my study, the University has filtered out about 50 % of the university entrants. Why? Because we have too many aspiring engineers? No, because the school has failed to give the future students a self-assessment. I mean with this, the undergraduates do not know what are their talents; they often start their post-school life without orientation. Too many undergraduates start a technical study although they would have better chosen something else. The contrary exists of course too.

This problem has as consequence of an incredible waste of time and money for the Industry because a lot of people are working in jobs, which do not correspond to them. I assert that one of two people work in the wrong job and study the wrong discipline.

### 5.5. Problems in imparting knowledge

I want to say some words about the pedagogical capabilities of university teachers. Some are excellent educators, but some have problems in imparting knowledge to their students. Why did one of my professors never think about the reason why his lectures became more and more empty until he was working with 5% of the initial number of students? The reason is so easy to find: he was unable to convey technical knowledge.

My recommendation: select new University teachers taking into account their scientific and pedagogical capabilities with the same importance, not scientific and then far behind educational.

A teacher in school has to learn much about how to educate his students. But a professor is by definition a talent in educating without learning.

## 6. OTHER SKILLS TO BE LEARNED AT THE UNIVERSITY

Besides the technical knowledge, which is the most important skill, there are other skills to learn at the University.

### 6.1. International Experience

You should perfection foreign languages, which you have learned at school. You should get in touch with foreign cultures.

In the past, in Central Europe, apprentices journeyed about acquiring experience from different masters (in German it is called «Balz»). When they had finished, they had obtained a package of experience, which they never would have gotten had they stayed all the time in the same place. Or: in the past, it was a sign of excellence to have studied at different universities. When you could say that you have studied in Göttingen, Paris and Harvard you received the highest respect.

We should return to this tradition: Europe especially is a perfect place where you can move around and learn about, and live in different cultures. It opens your horizons in such an important way. You will get experience for your entire life. The Universities should discuss together the possibility to offer a study, which would be done in different places, for example in Germany, Switzerland, France and the US. If I would have had the chance to undertake such an international study ; I would not have hesitated to participate.

### 6.2. Practical Experience

When I hire today new engineers, I find a lack of practical experience. This means, a lot of engineer jobs require basic manual capabilities, which often do not exist. Some engineers do not know in which direction they have to turn to remove a screw.

That is an unacceptable situation. Each engineer should have a minimum knowledge of basic, manual, technical matters. A short 4 weeks practice in a workshop before you start your study would be very effective.

*6.3. Professional Experience*

Professional experience can be already acquired during the University study. I mean with this, visits to producing enterprises. It should be an obligation for the students to have visited a certain number of enterprises. These 1-day-visits should contain a presentation of the company, of its activities, products and of it's manufacturing. Here, the student should get the possibility to learn how the manufacturing works. The corporate people should be able to respond to any question from the students. These visits will give the student an idea how and what is made? In addition the student gets an idea about his future job and sees which orientation and which type of product or activity is the best for him. The positive facets of such a program are multiple and are until today, underestimated. The more a student will see, the more he becomes ready for the industry. A student can learn in such 1-day-visits more than in four weeks of theoretical study. Industry and University have to work together. The industry has of course an interest to invest time in such student visits. My recommendation is to spend one day per week or per month for such company visits.

*6.4. Economic Basic Knowledge*

One of the most discussed questions would be: Has the Technical University to teach economic skills? My first response would be: how can we add new lectures to a completely full study program? It cannot be the goal to teach high-level economic knowledge besides the technical.

But: a basic knowledge would be more than helpful in the start phase of a young entrepreneur. What is a business plan? What is a balance sheet? What means depreciation? An engineer should be able to answer these questions at the end of his study. He should know, what is the structure of a company, which forms of companies exist, he should understand the logic of bankers and investors and he should know how function the economy. And not to forget: what can I do if I have an idea, with which I want to found a company? Each student should get a «manual», which explains in details the possibilities to start locally an industrial activity.

During my study I had the lecture in science of work and science of company organization. This lecture was the worst I had in my whole academic time. Purely learning of definitions by heart. Completely useless, I have not retained one thing from these lectures. Instead of this waste of time, a lecture in economic basics would be very good. A lecture full of real examples of the industry given by industry managers.

*6.5. Health*

Health is something to which the students pay less and less attention. The University sport activities in Europe are too much less supported as compared to the US. An entrepreneur has to have a good and stable health, otherwise he will become sick quickly with all the stress. The non-individual sports help in addition to learning social capability.

## 7. THE ROLE OF ENTREPRENEURS IN OUR SOCIETY

Still today some people (politicians, educators, managers) have not yet understood that in the case of growing or high unemployment, the only real, efficient way to attack this problem is to invest in new business. Small companies with a good business idea can quickly grow up and create stable employment. This has been completely forgotten for a long time. Only recently, since 2 or 3 years, the Europeans have discovered that in supporting start-ups, they can solve one of our biggest problems at this time.

But: first we have to sow before we can harvest the crop. Before we can take the benefit of our today's effort, perhaps ten years will pass. Therefore, lets not wait, help young company founders everywhere we can. In this way we can create «Blühende Landschaften» (translated: flowering landscapes or better roaring economy).

Everybody can contribute: the politicians have to help with governmental funding for start-up companies (excellent example: «KTI-Start-up» program of the Swiss government). They have to create legal conditions, which favorize the founding of new companies or venture capital investments. The politicians can create the framework to simplify the creation of new enterprises.

Industry can help with a lot of experience. Why not place a well-experienced industrial manager at the young entrepreneurs disposal. They could learn a lot from them, for example how to avoid making big, costly mistakes.

## 8. CONTRIBUTION OF THE UNIVERSITY

The Universities can:
- Modify the curriculum so that it is relevant to today's industry (general study, more methods and less definitions)
- Improve the pedagogic methods
- Shorten the study length, change the exam rules
- Require a short practice for manual skills
- Integrate visits to companies in both manufacturing and service sectors on a regular basis
- Promote the health of the students by offering a better sport program
- Teach basic economic know-how by presenting real cases of the industry

## 9. FINAL REMARK

I have given from my experience the skills, which are necessary to become a successful global entrepreneur as engineer. I wanted to make clear that there are things at the University not as they should be. Perhaps I could give an impulse in the modernization of the University. My hope is that at the end of this conference we will not leave Aachen and say: a nice meeting and then «business as usual» or «University as usual».

*Bernold Richerzhagen*
*Synova SA*
*Lausanne, Switzerland*

M. PÜTTNER

# STUDENTS' INTERNATIONAL PERSPECTIVE

**Abstract.** In a world of growing internationalization and globalization of markets, engineering education should not only consist of providing the essential technical skills.
The education's focus needs to take into account international working environments and project works, vast information networks across many different cultures to prepare the student engineer for the demands of the industry.
Projects span the globe, networks enable and demand close teamwork with engineers from all kinds of different cultural backgrounds. Languages represent the basic skills to enable communication and interaction, but only the far-reaching knowledge of a different culture permits mutual understanding.
This presentation is designed to encourage further expansion of the educational international network which is only at its beginning.

## 1. INTRODUCTION

In the light of how students gather information, what they may expect of the educational institution, which opportunities they have and what kinds of problems they encounter once decided to go abroad, the reader may notice existing deficits and help to improve international education.
This presentation is split into 6 chapters elucidating the following points:

## 2. OPPORTUNITIES

The opportunities to leave for another country are extensive and can be combined effectively with the demands of the educational system.

### 2.1. High School

The first important choice is to be made relatively early in High School regarding the future participation in exchange programmes. It is the selection of a foreign language as early as possible. The long-term memory and ease with which one can communicate in a foreign language depend more on the overall time exposed to this language than how intense the course actually was.

The second important choice is the best time to leave for a year abroad. Together with the curriculum of the exchange school this choice defines whether the exchange year abroad can be substituted for the school year at home. E.g. in Germany the best choice would be the $11^{th}$ grade, just before the last two years needed to qualify for the Abitur.

Since most High Schools do not raise enough their students' awareness of existing exchange programmes to foreign countries or of the best choice of time to leave, parents play an important role to point this opportunity out to their kids, encourage and help them to gather relevant information.

The experience of a year abroad is the most intense in High School. The students have the possibility to live together with a host family as if it was their own. With this intimate support kids make friends fast and get a close-up of the country's mentality, important for mutual intercultural understanding and avoidance of prejudices.

Students have the opportunity – which will never again appear in their life with the same intensity – to learn together with their new home classmates about their history, social & cultural concerns, learn & practice the language in spoken as well as written form in such a sophisticated way as is appropriate for learning one's mother tongue. As curricula and teaching styles vary between different countries, students have to find out how they can take advantage of this different approach to teaching and learning. Interesting and valuable courses might be at their disposal which they would never have had the chance to elect in their home institution.

The students' independence at a relatively young age promotes their sense of responsibility and self-reliance.

No other foreign experience will lead to the same level of language proficiency in the same amount of time.

## 2.2. Internships

Internships represent a valuable option before, during or after studies.

Important aspects of such a cross-cultural experience involve and challenge the whole personality of students who undertake an internship in a foreign country. It is the challenge to get personal and professional surroundings smoothly coordinated in a short period of time, exploring the situation and adapting to its various demands on the run.

However, students need strong support from their selected companies, especially with the preparations before the actual internship begins. Preparational tasks are much more difficult when performed from abroad, arising problems might be unsolvable without a helpful hand on spot. These tasks involve issuing documents necessary for visa regulations, arranging accomodation and assuring transportation.

During the internship the students are exposed to very varying backgrounds and differing mentalities of people working with them. This enables the students to improve their knowledge of individuals, how they think and how they act, in order to be able to react in an appropriate way. The improved knowledge of the foreign language and mentality prove valuable in international negotiations later on. Communication is facilitated and awkwardnesses can be avoided.

Moreover, the students obtain insights in working habits, methods of fabrication, quality assurance vs. deadlines, company policies and the difficulties in the interaction between co-workers and engineers or co-engineers and workers.

## 2.3. University

The opportunities to go abroad which are offered at university are abundant and include the whole spectrum of education as well as span the whole time period at the institution.

Lengths vary between several weeks to several months for study projects; academic exchange programs may take one or two semesters, double degree programs up to two years. Theses can be written or doctorates be realized at foreign institutions as well.

The international interest of professors at the home institution is mirrored in the diversity of opportunities. Common research of faculties all around the globe open up the way to study projects or extended research at the partner faculty abroad. Common projects between faculties and foreign companies not only promise applied studies but also direct implementation of obtained results.

Professors might not have current common projects with fellow-faculties or industry, but their wide-spun network nonetheless enables students to bridge the gap between continents. Without the necessary reference, foreign companies would rather stick with their own home students to avoid bureaucratic nuisance, difficult communication (time delay), health related issues etc.

The majority of students going abroad will take advantage of the official academic exchange programs offered at their university or of external exchange programs.

Pan-European exchange schemes such as SOCRATES are well developed and do not require explicit planning prior to 6 months before departure. Lists of equivalency of exams exist and facilitate the recognition of external exams.

Trans-continental exchange schemes such as FULBRIGHT require registration prior to more than 13 months before departure which unfortunately results in rigid planning of one's studies.

The opportunities to go abroad during studies prove to be the most diversified. Different approaches to learning and teaching at foreign academic institutions enables the student to value and criticize his home institution constructively. Different areas of academic interest or research help to broaden the students' academic horizon.

Language proficiency mostly improves as far as technical vocabulary is concerned, however, the overall language improvement is significantly minor to the improvement during the time at high school.

University mostly represents a cultural melting pot, the opportunity to meet people from all over the world helps to learn more about other cultures and mentalities. To live in mutual respect and friendship is enriching in a world where we are often confronted with racial discrimination and distrust.

## 3. OBSTACLES

*3.1. High-School*

Planning an exchange year abroad in high-school is the least problematic to countries like the USA. There are quite a few exchange organisations like ASSIST and AFS which take care of a lot of paperwork. They provide the necessary documents for students to obtain the exchange visitor status through a visa, select schools and/or guest families and provide local help in the foreign country when needed.

A major reason for students not to participate in such an exchange year can be put into four words: fear of the unknown.
- The fear to leave family and feel all alone and homesick,
- To leave friends behind, to lose them eventually due to one's long absence,
- Not to be able to make it back into one's former class and lose a year,
- To take the decision one year in advance, apply and sustain the thrive up to the actual departure,
- Of too much preparational work, dreaded paperwork.

These problems can arise quickly and grow in importance for the kids, thus it is necessary to help them explore their possibilities and at the same time giving them the impression that they can always fall back and rely on you, - as parent, friend or educational tutor.

*3.2. Internships*

There are ready available programs like IASTE to provide internships in foreign countries. Students apply for countries of their choice and after a selection process about one of every 3 students (numbers of 1998) is provided with an internship.
Local committees of IASTE in every country ensure that arising problems can be solved quickly.

This seems to be the least troublesome but also the least individual internship experience unlikely to fit concrete personal ideas.

Global companies offer international internships in various branches all around the globe, but mostly with the condition that the interested student performs an internship within this company in his mother country first.

Due to time constraints during studies it might not always be possible to fit in additional internships just to be offered the opportunity to spend the following one abroad. It also limits the possibility to get to know several different institutions.

The most individual and rewarding internship experience is guaranteed if students establish direct contact with companies abroad.

However, direct contact is mainly limited to telephone calls and correspondence, - depending on the distance between student and company a personal interview might not be an option. Due to this lack of personal contact face to face and a more general fear of arising medical, insurance, immigration, lodging and transport problems companies prove to be reluctant to accept foreign students.

If, however, this hurdle has been taken and the company accepted, many of these initially hypothetical problems actually come up and have to be dealt with quickly.

Speaking of personal experiences with an internship in the US: it took 10 months to get all the necessary documents from and redistribute them to 5 involved parties (university, US embassy, COUNCIL, German & US branch of the company).

Problematic impressions:
- US authorities announce wrong visa procedures & necessary documents
- without COUNCIL no legal background for visa exchange visitor status (payable)
- check position: internship compulsory to immatriculate, document of immatriculation vital for visa application procedures before the internship. (US afraid of German student staying in the US unless already enlisted at university in Germany)
- US authorities advising US company to reconsider internship and to take American applicant instead
- Transportation: excessive insurance policies on rental cars for age < 21, solution: buy car at arrival and resell just before departure,
- Car Insurances require US driver's licence within first 21 days of contract...

*3.3. University*

Exchanges between faculties seem to be relatively unproblematic in Aachen thanks to the department which is in charge of the university's international network, the Akademisches Auslandsamt.

Official pan-European exchange schemes, such as SOKRATES are well established, the application procedure is fast and unproblematic. Faculties have established a list of equivalent subjects to ensure exam credits for subjects taken at the foreign institution.

Projects can be arranged independently with responsible departments, possible credit can be discussed with the responsible professor.

However, 'exchange marketing' could be improved. Unless students have already discovered for themselves the value of international exchange and actively look for information, they will most likely be unaware of the chances they miss until the end of their studies.

## 4. ESTABLISH COMMUNICATION

The most accessible and up to date source of information is the internet.

Students look on company or university homepages to search for general information, opportunities, dates and deadlines, information about recent projects, opportunities to go abroad and / or to get a scholarship.

Company fairs on campus are the best events to establish direct contact with representatives of interesting companies. Individual questions can be asked, specific opportunities be discussed.

If the student has the possibility to obtain all necessary general information from the internet beforehand, such personal discussions lead to a much greater depth in the opportunities elucidated.

Personal contact within the university is vital as well, with the institutes – giving insights about current projects – or the Akademisches Auslandsamt, obtaining information from years of experience in the academic exchange process.

Student organisations such as BONDING provide important comprehensive databases.

## 5. GENERAL STUDENTS' VIEW

The subsequent graphs are taken from a study (1997) conducted by the Higher Education Information System HIS on behalf of the Federal Ministry of Education, Science, Research & Technology.

Figure 1 shows the students' perception of their own foreign language proficiency. As can be see from the graph, 43% have the opinion they do not have a good or very good knowledge of any language. Thus, the reading of technical or other specialized literature cannot be performed at a reasonable pace and there is no ease to communicate in a foreign language.

Considering that the vast majority of students had English for approximately 9 years in High School and a second foreign language for at least 3 years this is astonishing.

The need to maintain the students' high-school knowledge of foreign languages or even extend their knowledge is obvious. If students do not practice their language skills at university, they will lose their skills during their technical studies.

### Foreign Language Proficiency
*good / very good knowledge*

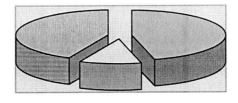

☐ *46% in ONE Language*
☐ *11% in TWO+ Languages*
☐ *43% in NO Language*

*Figure 1.*

Figure 2 relates to the students' view of necessary qualifications for employment.

According to the figures 'Qualifications for Working' the importance of work experience, English skills and social adeptness range around 30, on a scale from 0 (least important) to 100 (most important). Foreign experience is considered to be of negligeable importance to work later on with 9 points out of 100.

The question remains why the students do not esteem foreign experience, language skills and social adeptness valuable to an appropriate degree.

If job advertisements serve as an orientation for students, they get the message that work experience is the most important (70/100) and English skills important (55/100), without the emphasis of the importance of foreign experience (24/100) or social adeptness (31/100).

Industry should be aware of their role as opinion leader and pay attention to e.g. implied messages in job advertisements as they might have an impact on students opinion.

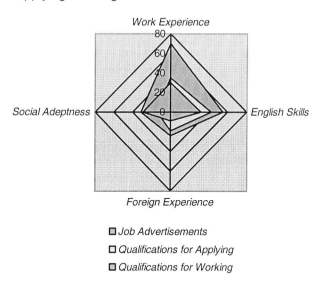

*Figure 2.*

Figure 3 portrays potential aid for seeking foreign experience.

Students seem to take not much advantage of optional opportunities and 30% state they need mandatory language courses to become active, 50% esteem it helpful.

Another 30% need strong urging of the institution to seek foreign experience.

Students seem to be lacking self-motivation to strengthen their language skills for a future participation in an exchange program and then to seek the foreign experience.

Language course offerings have to be improved in order to render them more attractive.

20% need adequate scholarship and 30% need job opportunities to be able to finance their stay in a foreign country.

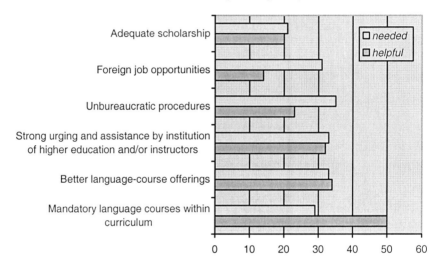

*Figure 3.*

## 6. CONCLUSION

Industry does not communicate its demand for internationally trained recruits early and strongly enough
- improve students' view on globalization and internationalization, importance of foreign language skills and international experience
- demonstrate open-mindedness

Students need challenge / aim to become active
- incorporate mandatory language qualifications

- incorporate international experience in the curriculum

More support
- financial support
- expand international network
- establish official international links
- improve links to national authorities

Raising the students' awareness of the necessity of an international education, improved preparation of students for the participation at exchange programs and further expansion of international exchange opportunities should be major goals for the future. Existing organizational deficits must be overcome to facilitate the growth of international networks (schools, universities, companies & authorities) in order to be prepared for growing internationalization and globalisation in the future. We are only at its beginning.

*Markus Püttner*
*Aachen University of Technology*
*Aachen, Germany*

# PART VI.

## SUCCESSFUL PRACTICE IN ENGINEERING EDUCATION -

### PROGRAMMES, CURRICULA AND EVALUATION

S. G. WALESH

# ENGINEERING THE FUTURE OF CIVIL ENGINEERING IN THE U.S.

**Abstract.** Traditional pride of U.S. civil engineers in their work increasingly competes with negatives. Examples of the latter are growing frustration with lack of appreciation and respect, the "shortage" of civil engineers, low compensation, reduction in academic preparation, "slippage" relative to essentially all other professions, encroachment of other professions and entities into the traditional CE arena, and absence from the political arena and public media. This paper critiques the U.S. CE education-experience-licensure-certification-continuing professional development process. It advocates a new paradigm consistent with the essentiality and complexity of infrastructure development and environmental protection in society. Civil works will always be in demand—that is unquestionable. To be decided, however, is who will lead the planning, design, construction and operation of civil works and protection of the environment; civil engineers or others? U.S. civil engineers are faced with a leadership and management challenge. They must engineer their future or others will engineer it for them.

## 1. INTRODUCTION

*1.1. Purposes*

This paper has three purposes, all of which apply primarily to civil engineering (CE) in the U.S. The first is to *celebrate CE's crucial historic, current, and future role* in protecting the environment and creating the infrastructure for civilized society. The second is to *critique the static, narrow focus of the CE education-experience-licensure-certification-continuing professional development process.* The third purpose is to *suggest a new paradigm for the CE* education through practice process, so that civil engineers can proactively embrace the future.

*1.2. Pride and Frustration*

Strong, opposing feelings of pride and frustration prompted the writing of this paper. As a civil engineer, the author is proud of the CE's historic role in providing for many of the basic physical needs of society. With the possible exception of medicine, CE may be the most important profession. The author's frustration begins with the increasing frequency with which civil engineers claim that, as a profession, "we get no respect." Adding to the author's frustration are low compensation, the "shortage" of civil engineers, and the gradual reduction in credit hours required for the BSCE degree major. Other negatives are CE's slippage in the education-experience-licensing-certification-continuing professional development process relative to other professions and our virtual absence from the political area and public media.

## 2. CONCERN: EMPHASIS ON CIVIL ENGINEER'S IMAGE

U.S. engineers, including civil engineers, are increasingly concerned with image. CE's principal response to being unappreciated is mounting image-boosting programs (e.g., ASCE, 1998) to publicize what we do. While publicizing what we have done and now do might help, a much more fruitful strategy would be to enhance what we do and how we do it so as to increase our societal value (e.g., see Walesh, 1998b).

## 3. CONCERN: LOW COMPENSATION OF CIVIL ENGINEERS

When engineers claim they "get no respect," it means, in part, that they are dissatisfied with the level of their compensation relative to two benchmarks: 1) the critical nature of their environmental protection and infrastructure development efforts and 2) compensation received by members of other professions. Based largely on the surveys and studies (AAES, 1998 and 1999; Alexander, 1991; Farr, 2001): 1) Inflation-adjusted salaries for entering and experienced engineers in the U.S. have remained essentially static for at least a decade; 2) Engineering compensation falls below that of most other professions; and 3) CE compensation falls below that of essentially all engineering disciplines.

## 4. CONCERN: "SHORTAGE" OF CIVIL ENGINEERS

The engineering profession in the U.S. seems to revel in concern with current shortages and predicting future ones. Often these shortages are cast as being very serious; of crisis proportion. Compensation data, however, do not support the "shortage" claims. Of course, whether or not a shortage exists, the perception of one serves the interests of several groups especially when the "shortage" message is successfully communicated to college bound young people. U.S. CE employers may benefit from "cheap" labor, U.S. universities could benefit by having more tuition paying students, and U.S. based professional societies may benefit by the potential for more revenue generating members. But does a "shortage" message, especially if not true or exaggerated, serve the long term interests of the CE profession and, ultimately, the public?

We can't have it both ways. Either civil engineers are poorly compensated or they are in short supply. Or perhaps the supply and demand situation is in a state of flux and will soon settle down to an equilibrium condition of either over supply (low compensation) or shortage (high compensation).

For an indication of the oversupply of civil engineers in the U.S., observe the "low level" technician and/or clerical work increasingly being done by engineers in private and public sector engineering offices. For a thoughtful discussion of misuse of professional personnel, refer to Perkins (2000). Cappelli's recent study (Wharton, 2000) of the "shortage" of IT professionals may further illuminate the civil engineer "shortage." Cappelli concludes there is no shortage of IT personnel. Instead the appearance of shortage is the result of poor recruitment practices, inadequate prospective employee assessment, and inattentive personnel management practices.

The resulting job dissatisfaction causes unnecessary churning, that is, high turnover. Might inadequate recruitment and retention practices help to explain the "shortage" of civil engineers? Or, if you are still not convinced of the oversupply of civil engineers, consider the fierce, price-based competition among U.S. CE firms.

## 5. CONCERN: GRADUAL HISTORIC REDUCTION IN CREDIT HOURS REQUIRED FOR THE BSCE DEGREE

Becoming a civil engineer is getting easier as indicated, in part, by the continued movement to reduce course credit graduation requirements. Elliott (1998) surveyed U.S. CE programs to learn more about what he called the "course reduction movement." Based on 51 responses, he found a range of 120.0 to 147.5 semester credit hours (non-military maximum) with a mean, median, and mode of, respectively, 132.9, 133.3, and 136.0 hours. Elliot says that engineering has "...actually moved backward, from about 150 semester hours in the 1950's to an average of 133 today."

Russell et al. (2000) examined BSCE degree requirements over the past 75 years for 11 leading CE programs dispersed around the U.S. The authors concluded that "...not only has the number of credits comprising a BSCE degree gradually but significantly decreased, but the total credit hours of engineering content has significantly decreased in many of the nation's leading CE programs..."

## 6. CONCERN: SLIPPAGE IN THE EDUCATION-EXPERIENCE-LICENSING-CERTIFICATION-CONTINUING PROFESSIONAL DEVELOPMENT PROCESS RELATIVE TO OTHER PROFESSIONS

Law and medicine, two professions that some engineers use as image benchmarks, have established education - experience - licensing - certification - continuing professional development requirements that greatly exceed those of CE. Most other widely recognized professions also have expectations more stringent than are in effect for CE.

### 6.1. Post-High School Education

Consider post-high school education. All major U.S. professions, with the exception of CE and nursing, require more than four years of post-high school education as the basic education required for licensure and the practice at the professional level. Examples are accounting (5 years), architecture (5 years), dentistry (8 years), law (7 years), medicine (8 years), optometry (8 years), pharmacy (6 years), and veterinary medicine (8 years). Some of these professions recently increased basic education requirements (e.g., accounting, architecture, and pharmacy). Meanwhile, CE retains the 200 year old four-year education model.

## 6.2. Continuing Professional Development and Specialty Certification

Based on continuing education required to retain a license, CE falls below accounting, dentistry, law, medicine, optometry, pharmacy, and veterinary medicine. CE ranks below dentistry, medicine, pharmacy, and veterinary medicine in availability of specialty certification.

## 6.3. Observations

CE, as well as all other U.S. based engineering professions, is behind all other major professions in the overall education-experience-licensing-certification-continuing professional development arena. Furthermore, given engineering's essentially static position, while other professions have progressed, engineering has slipped further behind.

## 7. CONCERN: ABSENCE FROM POLITICAL ARENA AND PUBLIC MEDIA

Another frustration, in the context of an increasingly technological world, is the U.S. civil engineers' essential absence from the political arena and the public media. Civil engineers seem to increasingly do the bidding of others. Others decide and direct and civil engineers do. Civil engineers are also rarely found in the public media, that is, as writers and speakers. Notable exceptions are civil engineers Samuel Florman (e.g., *The Civilized Engineer*, 1987) and Henry Petroski (e.g., *To Engineer is Human*, 1985) who have very effectively written about engineering for the general public.

## 8. VISION

I can build on pride and defuse frustration by looking forward (Walesh, 2000). My vision of year 2010 CE graduates embodies high value, long-lived traditions of CE. Included are a commitment to and satisfaction in **serving the public** through provision of much of society's physical fabric. Other essential traditional traits are **technical competence** within CE and allied fields; a **pragmatic**, get-the-job done drive and highly **ethical** behavior. In addition, and here is where more emphasis on changes is needed, to graduate from a program designed to provide entry into the practice of CE, a candidate should:

1. Demonstrate **technical competence** in several areas of CE such as structures, geotechnical, environmental, transportation and water resources.

2. Acquire broad exposure to the **humanities and social sciences**, including global art, history, literature, political science, philosophy, and psychology.

3. Understand **business and management** fundamentals including marketing, finance, human resources, project management, and profit.

4. Possess skill with five forms of **communication**—listening, speaking, writing, graphics, and mathematics and, in a related matter, understand the realities of cultural differences and know how to deal with them.

5. Appreciate the **ethical framework** within which CE functions holding paramount the safety, health and welfare of the public.

6. Be able to quickly find **data and information**, regardless of its global origin, and effectively evaluate, organize and present it.

7. Recognize the need for continuous, career-long **learning** and **career planning** and have the knowledge and self-discipline to do both.

8. Understand the need to identify and address **non-technical factors**, such as safety, aesthetics, economics and finance, in the planning, design, construction and operation of structures, facilities and systems.

9. Function as either a member or leader of a heterogeneous **team** of technical and non-technical personnel.

10. Complete a **practice-oriented experience**, such as an internship or cooperative education program.

## 9. EDUCATION: ROOT OF AND SOLUTION TO THE PROBLEM?

*9.1. History*

Engineering education was established in the U.S. **parallel** to liberal arts education. In contrast, the education of most other professions was structured in **series** with liberal arts education. Formal engineering study in the U.S. was born as and continues to be in **parallel** with liberal arts education. In sharp contrast, most other professions position their formal education programs in **series** with a liberal arts, or at least a more general education. As stated by Kerr (1995), "Students interested in becoming engineers generally studied engineering *instead of* the liberal arts, not *in addition to* the liberal arts." They still do. But, must this always be the model? Might integration of engineering, liberal arts and business into a longer program be a more appropriate paradigm for the future?

*9.2. Significance*

Partly because current CE problems grow out of the formal education model chosen by their predecessors, today's civil engineers must solve those problems primarily through educational reform. A new CE education paradigm is needed in the U.S. CE should be viewed from both within and outside of the profession as an incubator for tomorrow's leaders. Leadership should be part of the CE paradigm, given the very important charge to CE—providing for the basic physical needs of society and for environmental protection—and given the intellectual gifts of some of the young people who enter the field.

## 9.3. Role of Educators

Ideally, engineering educators should take the lead in creating a new CE education paradigm—in changing the curriculum, and, more important, its tone in terms of broader expectations. After all, educators are closest to the students and future engineers, and preparation for the future is the educators' business. Based on my experience as a practitioner, consultant, engineering faculty member and dean, I am not optimistic that CE educators will enthusiastically lead the movement to a new education paradigm. Two major obstacles stand in the way—tenure and isolation.

A few confident and courageous U.S. CE faculty and CE departments will initiate dramatic changes. Hopefully these individuals and departments will persevere and motivate others. But the vast majority of U.S. CE departments will change only in reaction to strong, external forces—if they change at all.

On the positive side, some bright lights illuminate the gray CE education landscape. Examples are:

- Conclusion, based on the 1995 CE Education Conference (ASCE, 1995) that "Increasingly, there is belief that an additional period of study, recognized by a professional degree, is *required* before (or even in the early years of practice) entering practice."

- Approval by the ASCE Board of Direction, at the ASCE annual convention in October 1998, of the masters as the first professional degree policy statement. A task committee was formed by the Board and charged with developing a vision of full realization of the policy and a strategy for achieving the vision. The task committee will present its report in October 2001.

- Implementation by the Accreditation Board for Engineering and Technology (ABET), beginning in 2001, of new criteria for accrediting U.S. engineering programs (ABET, 1997). *Engineering Criteria 2000* places much more emphasis on establishing institution-specific educational objectives in both technical and non-technical areas followed by on-going evaluation and improvement.

- Introduction by some universities, in response to external criticism, of alternatives to the traditional, unchecked tenure system such as more meaningful periodic post-tenure reviews; multi-year rolling contracts; increased use of non-tenured part-time faculty and optional parallel non-tenured but higher paying tracks.

## 10. ROLE OF PRACTITIONERS

In the short run, maintaining the present U.S. CE education model is in the best interest of private and public sector employers. The four-year, technically focused and producer-oriented education paradigm assures an ample supply of low cost, immediately productive, technically capable employees.

In the meantime, the principals of U.S. engineering firms and the senior managers of government entities will continue to complain about the inadequacies of entry level and experienced civil engineers. They are poor communicators, can't manage projects profitably, lack marketing interest and/or skill, get bogged down in technical matters, fail to meet client expectations, lack visibility in the community, fail to understand global context, and have no business sense. Deficiencies like these are typically resolved by training programs, lateral promotions, demotions, and firings. Perhaps these remedial costs are greater than the higher compensation that would be warranted by hiring better-prepared graduates of longer, more comprehensive CE programs.

In the long run, the functional narrowness, limited vision and large number of graduates will lead to more commodization of services and more intense price-based competition between consulting firms. Continuation of today's CE education paradigm will also steer more of the "best and brightest" away from CE or cause them to leave once they are on board.

## 11. A NEW PARADIGM

Educators and practitioners must jointly develop an approach to CE education and practice what is compatible with the changing world of work. The new model should attract and retain a core of bright, young people who want to be managers and leaders, as well as producers, and help them prepare to achieve those aspirations. While curricular changes are needed, *creation of an appropriate CE paradigm also requires a fundamental change in the tone, that is, the manner in which CE is portrayed to engineering students and to young practicing engineers.* A dozen suggested changes are:

1. *Assemble more heterogeneous faculties* that always include leading-edge practitioners who regularly teach all or major parts of courses vertically through the curriculum. Guest lecturers won't cut it.

2. *Increase faculty accountability and creativity* by replacing or markedly modifying the tenure system.

3. *Provide much more time.* Five or six years is suggested. The longer program does not necessarily mean five or six years in residence because distance learning will increasingly be an effective mechanism for education and training (e.g., see Walesh, 1998a).

4. ***Vastly expand the use of existing on-campus resources.*** Integrate the talents and course material of supportive, carefully selected faculty in complementary academic programs such as business, communication, history, law, philosophy, political science and psychology.

5. ***Convey the essential historic, global role of the "civil engineer,"*** back to at least the beginning of recorded history, in meeting society's physical needs.

6. ***Instruct students in business fundamentals.*** Examples include finance, decision economics, human resources, law, ethics, the role and selection of consultants, and marketing.

7. ***Enhance communication competence.*** Effective communication—use of mathematics, graphics, writing, speaking and listening—is necessary, although not sufficient, to realizing one's potential in CE.

8. ***Expand teamwork instruction and experience*** with focus on projects. Teamwork is the norm in the practice and business of CE. Students need to understand that the complexity of most engineering projects requires fielding a team of specialists and generalists.

9. ***Require a meaningful paid cooperative education or internship*** experience. Create a mutually beneficial three-way partnership among the student, the engineering employer and the educational institution.

10. ***Provide instruction and experience in various ways of learning.*** Examples are self-study, live multistation audio-video instruction, Internet and networking. Given that much of what we now know has a short, useful half-life, knowing how to learn rather than only knowing will be increasingly valued.

11. ***Enable students and faculty to interact with exemplary civil engineers;*** leaders, managers and entrepreneurs. This is the role for guest lecturers. Encourage the guests to describe the why, how and who aspects of major CE projects.

12. ***Recruit and hire graduate engineers who have earned masters degrees and have cooperative education experience.*** If technicians are needed, and they should be, hire graduates of technology programs and assign them to young engineer supervisors.

## 12. CONCLUDING THOUGHTS

Civil works will always be in demand—that is unquestionable. To be decided, however, is who will lead the planning, design, construction and operation of civil works. Civil engineers or others? Our environment will increasingly need protection. Civil engineers could lead this effort, but will they? Civil engineers can engineer their future or others will engineer it for them. "The world is run by those who show up" (Weingardt, 1997). If we truly want to change the CE paradigm, to be more "on top" and less "on tap," we must head in a new direction.

> Two roads diverged in a yellow wood,
> And sorry I could not travel both…
> I shall be telling this with a sigh
> Somewhere ages and ages hence:
> Two roads diverged in a wood, and I—
> I took the one less traveled by,
> And that has made all the difference.
> (Robert Frost)

## 13. REFERENCES

Accreditation Board for Engineering and Technology (1997). "Engineering Criteria 2000," Third Edition, December.

Alexander, J. A. (1991). "Professionalism and Marketing of Civil Engineering Profession," Journal of Professional Issues in Engineering Education and Practice – ASCE, Vol. 117, No. 1, January, pp. 10-20.

American Association of Engineering Societies, Engineering Workforce Commission, Professional Income of Engineers:1998.

American Association of Engineering Societies, Engineering Workforce Commission. (1999). Engineers, October, 8 pages.

American Society of Civil Engineers (1995). "Summary Report: 1995 Civil Engineering Education Conference."

Elliot, R. P. (1995). "Crisis in Engineering Education: Survey of Civil Engineering Curriculums," Forum, Journal of Professional Issues in Engineering Education and Practice – ASCE, Vol. 124, No. 1, January, pp. 8-10.

Farr, J. V. (2001). "Commodities and Value Based Pricing of Engineering Services," draft paper (For current version of the paper and other matters, contact the author at jfarr@stevens-tech.edu.)

Florman, S. C. (1987). The Civilized Engineer, New York: St. Martins Press.

Kerr, O. S. (1995). "The Economics and History of Engineering Education," The Bent of Tau Beta Pi, Fall 1995, pp. 9-11.

Perkins, D. (2000). "The Misused Project Manager," PM Network, Project Management Institute, Vol. 14, No. 9, September 2000, pp. 88-93.

Petroski, H. (1985). To Engineer Is Human, New York: St. Martins Press.

Russell, J. S., B. Stouffer and S. G. Walesh. (2000). "The First Professional Degree: A Historic Opportunity," Journal of Professional Issues in Engineering Education and Practice – ASCE, Vol. 126, No. 2, April, pp. 54-63.

Walesh, S. G. (1998a). "Corporate Universities: The Concept and Its Potential Application Within Civil Engineering Organizations," presented at the session "Corporate Universities in Civil Engineering Organizations," Annual Convention and Exposition, American Society of Civil Engineers, Boston, MA, October.

Walesh, S. G. (1998b). "Short on Image, Short on Substance," Engineering Times, December.

Walesh, S. G. (2000). "Engineering a New Education," Journal of Management in Engineering – ASCE, Vol. 16, No. 2, March/April 2000, pp. 35-41.

Weingardt, P. (1997). "Leadership: The World Is Run by Those Who Show Up," Journal of Management in Engineering – ASCE, Vol. 13, No. 4, July/August, pp. 61-66.

Wharton School (2000). "Are Tech Workers in Short Supply?" University of Pennsylvania, CNET News.com, September 28.

*Stuart G. Walesh*
*Consultant*
*Valparaiso, Indiana, USA*

J. HAGENAUER, J. BARROS, C. BETTSTETTER, S. JAUCK

# THREE YEARS OF EXPERIENCE WITH AN INTERNATIONAL GRADUATE PROGRAM AT TU MÜNCHEN

**Abstract.** The Technische Universität München (TUM) offers a new high-level graduate program that results in the degree Master of Science in Communications Engineering (MSCE). This pilot program is designed for international students and, thus, is taught entirely in English. It has a duration of two years, including a ten-week internship in a German company and a six-month research period for the completion of the master thesis. In this contribution, we will show the results of a three year experience with this program, focusing on recruiting and selection procedures, international university marketing, flexible management, cooperation with industry partners and multicultural learning environments. At the moment 57 students from 20 different countries participate in the program. We believe that our MSCE graduates – being acquainted with the German culture and having enjoyed a close relationship with German industry in this field – are well equipped for leading positions in globally operating companies in their home countries and worldwide.

## 1. INTRODUCTION

In an effort to increase the study opportunities for foreign students and, thus, promote the internationalisation of the Technische Universität München (TUM), the Department of Electrical Engineering and Information Technology started in October, 1998, an international graduate program, taught completely in English language and leading to the degree "Master of Science in Communications Engineering" (MSCE). Since then, three classes, each of about 20 to 35 students from 28 countries all over the world, have come to Munich for their graduate studies. During the first three years of experience with this new program and its innovative concept, both the students and the department's staff went through an extensive learning process, influencing the course organization, the teaching methodologies, the cooperation with the industry, and, of course, all aspects concerning the department's increasingly multicultural environment.

After a brief presentation of the MSCE Program, we will discuss some of the choices made in terms of the program's management, as well as recruiting and international marketing. A certain emphasis will be placed on the relations with the industry sponsors, as they play a key role in the success of this endeavour. Finally, we will focus on intercultural communication and describe the international learning environment. We hope that the insight and the motivation gained during the first three years of experience with the MSCE Program will serve as encouragement and guidance to colleagues willing to go for a similar experience.

## 2. THE MSCE PROGRAM

The MSCE program, planned in 1997/98, was TUM's first step towards the international higher education market. The pedagogic aim of the MSCE Program is to provide the student with a solid background in the theory and fundamental concepts of communications engineering, insight into current trends and developments of the field, as well as ample opportunities for practical experience. For this to be accomplished, the program has a duration of two years, divided into four semesters, including a ten-week period for the completion of the master's thesis.

### 2.1. Course Organization

The academic part of the program, which begins every year in October, consists of 10 mandatory courses and 3 out of 8 possible elective courses. In addition, the student must choose 2 out of 7 labs and attend a seminar, where he must prepare a topic and give a presentation. The last semester is fully dedicated to the completion of the master's thesis, which constitutes the final requirement for graduation.

### 2.2. Teaching Staff and Facilities

The Department of Electrical Engineering and Information Technology (EI) has 1500 students, 20 institutes, 32 full and associate professors and 250 teaching and research assistants. The majority of the courses are held by the department's full professors, most of whom are IEEE fellows. In the summer semester, well-known international professors take over two courses. The lectures of the MSCE program are held in classes of no more than 40 students. In labs and semester projects, students are assisted individually by staff members. Each students is provided with office space, a computer connected to the Internet and access to university libraries. During the semester and the spring break, German courses are available at the TUM language centre.

### 2.3. Course Fee and Financial Support

The current regular tuition is US$ 6000 per year. However, grants from industry are available, covering the tuition plus a paid internship with the sponsor company. So far, all admitted students have received a tuition grant, sponsored by the department together with its industry partners. Further grants covering part of the living expenses or travelling expenses can be obtained through the German Academic Exchange Service (DAAD) or some of the sponsor companies and other organizations.

## 3. INTERNATIONAL MARKETING AND RECRUITING

For an international graduate program to succeed, it is essential to recruit very good students from top universities abroad. This is not an easy task, especially considering that American and British universities hold a well established and extremely dominant position in the international education market, as opposed to German universities, whose marketing offensives have only recently began to make an impact in Asia and South America. One of the major barriers for foreign technical students to come to Germany has always been the necessity to learn the German language. By offering courses in English, it is possible to render German universities more attractive to international students. Furthermore, it is important to emphasize the major advantages that arise from the close association between German universities of technology and well-known multinational companies, like Siemens, Infineon Technologies, Ericsson, Rhode & Schwarz or BMW. Since these are companies with strong investments abroad, they constitute an important reference for foreign students, as the latter consider not only the academic aspect of their graduate education but also their future career prospects. In countries, where most graduates currently go to the United States fur further studies, a European experience clearly becomes a competitive advantage in the employment market.

*3.1. Admission Requirements*

To be considered for admission, candidates must hold a bachelor's degree (BSc) or an equivalent degree in electrical engineering, information technology, computer science or in a related field from a well-known university or college. Since all teaching is done in English, non-native speakers of English are required to supply adequate TOEFL results. A modest knowledge of German is only required for everyday life.

*3.2. Application Procedure*

In order to apply, potential candidates can download the application form from the MSCE homepage on the Internet (http://www.master.ei.tum.de) or send an email to the MSCE Admission Office (master@ei.tum.de). The application form includes a personal history, a recent photograph, a list of academic and professional activities, as well as short essays on previous work and motivation for further studies. In addition the candidate must provide certified photocopies of academic records, a notarised translation into English, German or French, a TOEFL certificate and two recommendation letters from professors or former employers. The latter should indicate how the recommended candidate ranks among his or her peers.

*3.3. Selection Process*

After the application deadline, all the submitted documents are analysed by the board of examiners, which makes the first selection. The selected candidates are then interviewed in English, mostly via telephone by one of the professors, who tries to evaluate not only the candidate's communication skills, but also his knowledge on a number of basic concepts which are essential for the pursuit of graduate studies in communications engineering. The final decision is made based on the previous analysis and the recommendation of the interviewer. This selection procedure has proved very effective, since it allows us to evaluate to a certain extent how well the candidate is qualified for the program and whether her or she will be able to adapt to a different country and a different education system. Furthermore, it offers a complementary indication on the English level of the candidate, since oral and technical language skills are not evaluated by the TOEFL.

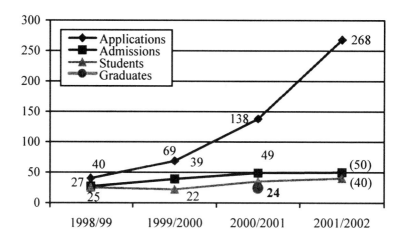

*Figure 1. Recruiting statistics for the MSCE Program.*

*3.4. Marketing and Recruiting*

In pursuit of the aforementioned objectives, an international marketing strategy was laid out which privileges the personal contact to key academic partners in top universities, mostly in Asia, South America and the Middle East. Through regular visits and alumni efforts, it was possible to quickly establish a network, which efficiently conveys the information about the MSCE Program to prospective students.

*Table 1. Distribution of the first three classes (1998-2000, 1999-2001 and 2000-2002) according to country of origin*

| Country | Number of Students |
|---|---|
| Australia | 1 |
| Bangladesh | 1 |
| Brazil | 4 |
| China | 18 |
| Greece | 7 |
| Iceland | 1 |
| India | 7 |
| Indonesia | 5 |
| Iran | 1 |
| Japan | 1 |
| Korea | 2 |
| Lebanon | 9 |
| Malta | 1 |
| Mexico | 2 |
| Morocco | 1 |
| Pakistan | 1 |
| Poland | 1 |
| Portugal | 1 |
| Serbia | 1 |
| Slovakia | 1 |
| Sudan | 1 |
| Taiwan | 1 |
| Turkey | 2 |
| UK | 1 |
| USA | 1 |
| Vietnam | 3 |

Six months before the application deadline, our partner universities receive a complete information package, containing leaflets, posters and the most recent yearbook edition. A lot of effort has also been put in a strong presence in the Internet through an informative and appealing homepage, which includes an online version of the yearbook. We hope to limit our administrative workload by introducing an online application system, which is currently being tested. The attendance of international education fairs around the world helps to keep track of the latest trends and developments in the international education market. The effectiveness of this marketing strategy is well illustrated by the recruiting statistics shown in Figure 1. The number of applications has continuously increased and the admission-to-application ratio is now approximately 1:5. Although the expected value of the number of students is 35 per year, since the best applicants generally apply to several other universities, 70 candidates are selected for interview and about 50 of them are finally admitted to the program. Table 1 shows the countries of origin of the MSCE students in the period of 1998-2001. Notice that about 80% are non-European and more than 50% come from Asian countries.

## 4. MANAGEMENT AND COORDINATION

The experimental character of the MSCE program also allowed for the trial of new concepts and ideas in relation to the program's management and coordination. Once the regulatory procedures were met by the administration of the university and approved by the Bavarian ministry, a high level of autonomy was granted to the MSCE program, including the freedom to pursue its own strategy, to choose its students according to self-defined criteria and to manage the self acquired sponsoring funds. By avoiding excessive bureaucracy and centralized decision-making, this proved to be a very efficient way to guarantee the necessary flexibility to a program which was just beginning.

The decisions concerning the strategy of the program, the choice of its students, and all the pedagogic aspects are made by a board of six professors, which is headed by the program director. The latter is aided by the assistant program manager who runs the daily business, the international marketing and most of the student counselling. An administrative assistant takes care of the many inquiries and application forms, while a student aid is responsible for the constant actualisation of the homepage content. This small team is able to cope with the high workload associated with the MSCE program in a very effective way.

It should be noted that the university only supplied the position of the assistant program manager and a small start-up capital. All the other human resources (professors, assistants, lab tutors) are guaranteed by the chairs of the electrical engineering department.

## 5. SPONSORSHIP AND INDUSTRY COOPERATION

The interaction with the sponsoring companies is manifold and very enriching both for the university and for our industry partners. Due to the high demand of young, qualified engineers with an international background, it is possible to find sponsors who pay for the tuition fees in return for a close contact to the students during their graduate studies. To avoid unnecessary strains, no direct correspondence is made between the sponsor company and the sponsored student. Thus, company information is conveyed to all the students, and there is no need for an official commitment from the students' side. During the paid internship, which is mandatory, the students have the chance to acquire some practical experience and get to know the industry more closely.

Another aspect of this cooperation lies in the collaboration of invited adjunct professors, who are currently working in the industry and come to the university to teach the more practically oriented courses. In addition, since all of the full professors in the department have themselves spent several years in a company before being appointed to the university, the students benefit very much from their practical experience and their professional contacts.

## 6. INTERCULTURAL ASPECTS

Given the international character of our graduate program, it is important to mention some of the challenges posed by the resulting multicultural environment. When the first MSCE students arrived, a lot of effort was put in helping them to adjust to a new country and a different culture. Since most of them did not have German skills, it was essential to guarantee their accommodation before their arrival. This was done with the aid of the student service ("Studentenwerk"), which gave them small apartments in the former Olympic housing complex. In addition, an airport pick-up service was organized and tutors guided the students in all steps from the registration at the city office to the matriculation at the university, including health security procedures and getting acquainted with Munich.

This additional tutoring caused some concern among German diploma students, who felt that the new foreign students were being privileged in terms of counselling. Unfortunately, mainly because of the strict classroom system, the contact between the master and the diploma students is still scarce, and this constitutes one of the major concerns of the MSCE board. In the hope to improve their relationship, a joint lecture takes place in one of the semesters and some social activities are organized ranging from soccer games to excursions and student parties.

For the professors and their lecture assistants the multicultural classroom also put a series of challenges. For the first time, teaching in English language became part of the university's everyday life and some pedagogic changes were introduced in the system, like graded homework and midterm exams, reflecting some of the methodologies proved successful in universities in the United States. These new activities build an experimental framework for the department's creation of BSc and MSc programs for all students.

Coming from many different countries with different cultures and higher education systems, the students showed very distinct reactions to the teaching and organization practice in a German university. Typically, an Asian student expects more counselling and more frequent advice than the average European student. While South American and Indian students adapt well and quickly try and socialize with all the others in the classroom, Arabic and Chinese students tend to stick to their group of countrymen, with whom they can speak their own language and share their way of life. Still, the experience shows that during the two years of the program firm cultural bridges are established between the students and the wealth in intercultural communication became major asset both for them and for the university. We also noted that after two years the graduates acquired a good command of the German language and the large majority of them integrated very well in Munich's social and cultural life.

## 7. CONCLUDING REMARKS

We have described the introduction of a new international graduate program at the Technische Universität München and discussed the challenges and the benefits of this innovative step. A well-defined set of objectives and a highly motivated teaching staff, aided by fruitful cooperation with the industry, adequate marketing and flexible management, led to the large success of the MSCE program.

The model we presented is now being followed by other departments at the TU München, thus securing our university's international position. For our part, the starting phase of the MSCE Program has been very enriching both from a professional and a personal point of view. In fact, it is very rewarding to know, that 22 of our first 24 graduates remained in Germany: 4 began their PhD studies at the TU München and 18 started their professional career in German companies.

*Joachim Hagenauer, João Barros, Christian Bettstetter, Silke Jauck*
*Technische Universität München*
*Department of Electrical Engineering and Information Technology*
*München, Germany*

H. DE RUITER

# FUTURE DEVELOPMENTS OF THE EUROPEAN MINERAL PROGRAMS

**Abstract.** In 1996 Delft University of Technology took the initiative to develop a joint curriculum for the fourth year's Mining students, the "European Mining Course" (EMC). Fourth year students of four participating universities form one group during an eight months period. This group of 15-20 students stays two months in Helsinki, London, Aachen and Delft to follow a joint curriculum (in the English language). This enables each of the universities to concentrate on their strong fields of expertise. The success of the EMC resulted in the start of a second program in September 1998, the European Mineral Engineering Course (EMEC). The program includes courses in mineral processing, metallurgy and recycling. Both programs have students from the four organising countries as well as exchange students from Canada, Australia, Argentina, Chile, Poland and Spain.
Industry supports this unique initiative and participates in an organisation: the "Foundation of European Mineral Engineering Programs" (FEMP) which gives financial support to the students participating in the programs.

## 1. INTRODUCTION

The current situation in Europe is that globalisation of the raw materials supply has caused the larger mining and the supply companies to expand and invest particularly outside Europe. These companies still account for a fair percentage of world mineral production, (over 20% of some minerals), from their mines inside and outside of Europe. It is also estimated that European-based companies account for 20% of worldwide mineral exploration (of which 10% takes place within Europe). Therefore these industries still have significance for the European Union and its members. In addition it should also be emphasised that the E.U. is a substantial consumer and processor of imported minerals.

All of this means, beyond doubt, that mineral engineering is still important in the E.U. It is therefore vital to maintain the very high level of knowledge and skill base in mining engineering and mineral technology that currently exists in the E.U., through a good and well-balanced education program.

The changing circumstances, since the seventies, led to new challenges in the organisation of the education at the European universities in the field of mining engineering and minerals engineering, the latter including processing, metallurgy and recycling.

Those accepting the challenge adapted to the situation by renewing the curriculum and looking for ways to include the modern requirements in their programs like environmental engineering, economical subjects and postgraduate courses. The global market for engineers also led to more exchange of students between universities. This ensures that Europe will be able to maintain the skills in this important economic area.

## 2. EDUCATION

At the end of the eighties Delft University of Technology realised that it was necessary to look for innovative ways to continue, economically, a high quality mining program. It was then decided to send the mining students during their final year to Imperial College's Royal School of Mines in London. This was the first initiative of this kind in Europe en led in 1995 to further developments.

Early 1996 discussions took place between Delft, London, Helsinki (Finland), Aachen (Germany) and Luleå (Sweden These universities were primarily chosen because of existing links, similarity in structure, culture and the wish to further internationalise.

The first four of these universities took the decision to establish a joint curriculum during the final year. The general thought was that each partner would concentrate on subjects in which it was already strong. Thus a joint, high quality, curriculum was realised. The concurrent effect was that each university did not have to maintain the entire curriculum, which in turn yielded cost reductions.

Undergraduate students spend the first part, the bulk, of their university studies at their home university. There they obtain the basic engineering skills and a fundamental understanding of mineral science and engineering. However, during the fourth year, before commencing with the M.Sc. thesis work, the students of all 4 universities spend, as a group, two months at each of the four universities to study the various specialisations.

Immediately following the courses the students choose to develop special expertise in one or another aspect of mining or mineral engineering and complete a thesis project. Upon completion the home university continues to award the M.Sc. diploma.

Since that time two educational programs have been developed as one-year co-operative Masters programs in mining and mineral engineering. In this co-operation between the technical universities in Helsinki, Aachen, London and Delft students spend some time at each institute for a total of 8 months during their 4th year. The European Mining Course (EMC) program started as the first one in the academic year of 1996/97 with eight participants. The results were very encouraging and it was decided to increase the number of participants to about 20 - 25 maximum. In 2000, the program is running for the fifth time in its full form. Since 1998 a similar program has been developed in the field of minerals processing / extractive metallurgy / recycling: the European Mineral Engineering Course (EMEC).

Total enrolment of students from the four organising universities has so far been below the maximum capacity of the programs. The extra room has been made available to students from universities having an exchange agreement with one of the four partners. So far students from the following universities have joined one of the programs:

| | |
|---|---|
| Queens University (Canada) | University of Queensland (Australia) |
| University of Wroclaw (Poland) | University of Madrid (Spain) |
| University of Leoben (Austria) | University of Freiberg (Germany) |
| University of Clausthal (Germany) | University of Concepcion (Chile) |
| University of San Juan (Argentina) | Virginia Tech. (USA) |

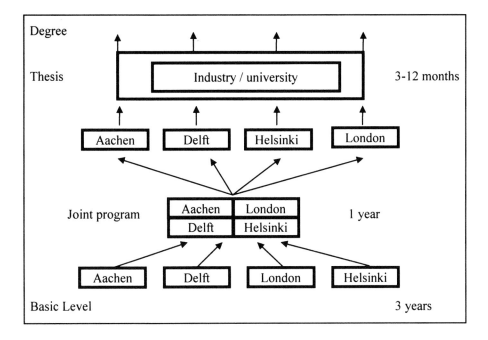

*Figure 1. European mineral engineering course (EMEC).*

## 3. ORGANISATIONAL AND FINANCIAL ASPECTS

The course director (TU Delft) takes care of the overall organisational aspects of the programs. These include:
- registration (and selection) of the participants;
- maintaining the Course Calendar, the Course Program and the Course Descriptions;
- administer the ECTS credit points and the official certificates;
- production of a Course Summary at the end of the academic year;
- keeping the Web pages up to date;
- maintaining contact with educational institutions, industry and alumni;
- taking care of administrative and financial matters;
- public relations matters.

The cost of giving the courses is in fact part of the universities normal cost for education. Combining the curricula now reduces the cost to an economical level, as the partner universities do take care of part of the education.

The financial needs for students (in addition to what they normally need during the year) amount on average to 2,500 Euro per student, of which up to 1,000 Euro has been covered by travel bursaries from the European Commission (Erasmus / Socrates programme).

The additional costs to the universities are mainly travel and accommodation expenses for company visits and are estimated at 5,000 - 10,000 Euro per university per year.

Ideally most of these funds will ultimately be generated through the fees from outside participants. Until that time an appeal has been made to Industry for financial assistance.

## 4. FEDERATION OF EUROPEAN MINERAL PROGRAMS (FEMP) AND INDUSTRY

On 16 December 1999 the Federation of European Mineral Programs (FEMP) has been established which acts as an umbrella over the various current and future programs. FEMP will a/o set guidelines for standards to participate in common educational activities and take care of common financial matters.

This *Federation of European Mineral Programs (FEMP)* was established on 16 December 1999 in Delft. The two main goals are

- ➢ Establishing stronger ties between the partners and opening up possibilities for other (European) universities to get a formal involvement in the programs, through contracts with FEMP. Involvement could be through participating students, exchange of staff or being part of an Open and Distant Learning (ODL) program (courses via Internet). Developments for ODL will start in 2000. This will ultimately lead to a full scale integrated European network of mineral programs and will be a first in the world on that scale.

- ➢ The European mining and mineral industry, united in Euromines, has reacted favourably on the developments in the field of education. The Federation will be the formal contact between Industry and the European universities. Member companies of Euromines and a number of other companies have supported the programs from the beginning.

The Board of directors of the Federation includes professors and a representative of Euromines, who acts as a representative of the major European mining / mineral companies. Several other companies have made commitments, or expressed interest, to support FEMP. At this moment ABN-AMRO, Caterpillar and Pricewaterhouse Coopers are supporting FEMP. Representatives of these companies are members of the Industrial Advisory Committee.

This committee meets at least once a year with the FEMP board. Members are being consulted regularly for advice and co-operation.

Basis for co-operation with Industry is that there are joint benefits for the Industry and for the programs like the following:

*Publicity*
The name, logo and other company information will be mentioned in course material, etc.

*Recruitment opportunities*
Companies get opportunity to learn to know the participating students through involvement in the programs.

*Access to the educational network*
One of the participating universities can be contacted in case of special requests for information, expertise, manpower or other matters. The appropriate partner and associated universities will then also be contacted.

*Membership of the Industrial Advisory Board*
A representative of the participating companies will be member of the Industrial Advisory Board (IAB). Once a year FEMP organises a meeting between the FEMP and IAB boards. At this meeting various subjects are discussed like the course contents, project work, etc.

*Special courses*
Special courses can be arranged for companies. These courses will be based on the best expertise available from the educational network. Companies participating in FEMP will get a special rate.

*Participants from industry with the programs*
Companies, supporting FEMP, will get priority and a special rate if their staff would like to participate in one or more of the modules.

*Students for internships and special projects*
It is possible to use students to carry out work on special projects and studies.

## 5. FURTHER DEVELOPMENTS OF INTERNATIONALISATION

*Europe*
An increasing number of European universities have expressed interest to get some involvement with the EMC and the EMEC programs. So far the response has always been that EMC and EMEC partners do not want to be "exclusive" parties, but that they want to act as a core for future co-operation. Involvement of other European universities is in principle no problem, provided that the organisation remains practical (expanding the "travelling circus" with more locations is not really practical any more). Other solutions have to be explored.

In January and February 2000 discussions took place with Nancy (together with Euromines), Trondheim and Luleå. Besides these three universities, a few other universities have informally expressed their interest (a/o Ljubljana). It is also expected that there will be interest from some of the Southern European countries (e.g. Portugal, Spain, Greece) as well as from other Eastern European countries.

In order to expand the current exchange, other universities in Europe will have to develop similar activities. This can be done in several ways:

➢ Students from other European universities are attracted to the existing EMC and EMEC programs. Unfortunately there is only limited room for extra students in the programs.

➢ When the number of students exceeds the maximum, additional cluster(s) of universities will have to be formed. Clusters could be a combination of current partners and universities from other countries, or alternatively there could be clusters solely consisting of universities from other countries. On these additional circuits it is not necessary to teach exactly the same courses. Actually emphasis should be on other fields but still under the umbrella of the European Mineral Programs. At this moment discussions are being held between Leeds, Oviedo and Athens, as well as between Berlin and a number of Eastern European countries.

Greater international involvement will widen the base of knowledge within the programs and give greater opportunity to involve. Students will then be able to choose which cluster they prefer. Alternatively teaching staff can be exchanged. Following diagram gives an example of possibilities.

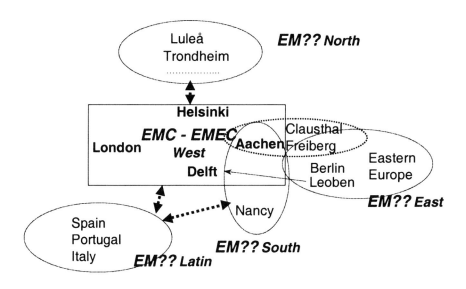

*Figure 2. Possible structures.*

Bottlenecks at this moment are some of the rules of the Socrates and other exchange programs, which make it difficult to exchange students to universities without agreements with all program partners. Ideally Socrates should have a possibility for universities to make an agreement with for instance FEMP or the EMC / EMEC programs. Another matter of concern is how to involve students from Eastern Europe. The amount of the Socrates mobility grants is by far not sufficient to cover all costs of travel and accommodation plus food.

*Further international developments*
Similar activities as in Europe could be developed in other countries like Canada, USA, Australia, etc. Australia has already taken initiatives in this direction. Other possibilities are the exchange of students between programs in different countries. To a certain extend this is being done with the EMC and EMEC, but additional opportunities are needed.

The following diagram shows a possible structure that could develop in the next years:

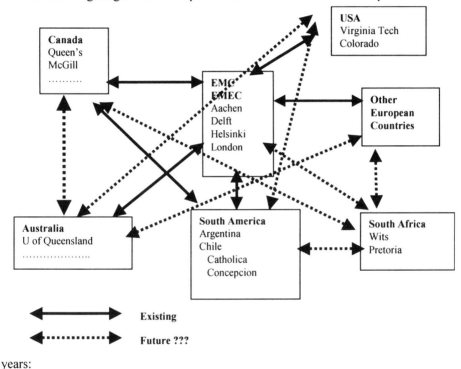

*Figure 3. Worldwide network.*

*Staff exchange*
There is also room for more movement of staff, particularly to other universities that wish to co-operate with FEMP. Staff from the core four universities could offer their courses as modules to the other universities in the periods when the EMC students are not with them. Alternatively staff of other universities could offer specialist courses to core universities.

*Open and distant learning techniques*
Another possibility is to develop Open and Distant Learning techniques. Using ODL the current courses will be made available to other universities, but O.D.L. can also be used to develop special courses. Universities can contribute to the program, e.g., by offering a single special course. This course can be included in the curriculum by means of ODL, without having to visit the institute physically. Development of courses can also be done jointly between two or more universities. The various courses can then be published on a webpage. At this moment TU delft uses a system called Blackboard. The possibility to expand this to partner universities is being studied.

*Exchange of teaching material*
A great similarity of contents of various courses at the different universities exists. Some of this material could be shared, e.g. one university develops and maintains course A and the other university develops and maintains course B. Both universities use the material of A and B. An example is TU Delft, which uses the "Introduction to Mining Course" of Queen's University.

*E- communication*
An electronic communication board on which universities can post questions regarding technical or other subjects could be beneficial. Compare this with the "Ventilation communication board", which is co-ordinated by Pierre Mouset-Jones. Questions dealing with education, research, internships, jobs, etc. could then easily be distributed.

*Hans de Ruiter*
*Delft University of Technology*
*Delft, The Netherlands*

H. RAKE

# TEACHING CONTROL ENGINEERING TO MECHANICAL ENGINEERING STUDENTS BY A COMBINATION OF TRADITIONAL AND MODERN METHODS

**Abstract.** Control Engineering is the only subject of the main study program that is mandatory for all students of mechanical engineering at RWTH Aachen. This has the consequence of between 600 and 1000 examinations annually and the duty of the teaching staff to work as efficiently as possible in transmitting the essentials of this subject. This efficiency has been obtained by a carefully balanced combination of traditional and modern methods of instruction.

## 1. INTRODUCTION

The essentials that have to be taken care of when teaching control engineering belong to two distinct areas. The first area is the basic understanding of processes within controlled devices and controllers of all kind. The second area is the knowledge of the rigorous mathematical ways and means to be used to analyse and design controllers and controlled systems. Both areas can be mastered only with a certain effort on the side of the teacher as well as on the side of the learner and the benefits of this effort can be enhanced by use of modern visual aids.

## 2. TRADITIONAL METHODS

The traditional basis of the course in control engineering is a course brochure of about 350 pages containing almost all facts and figures used and mentioned during the three hours of lectures given each winter semester. This brochure is updated and printed annually.

The subjects presented during the lectures are consolidated by examples which are treated in two hours of exercises. In one of these exercises problems are worked on with the aid of partial solutions. This gives the opportunity to treat realistic problems without getting lost in too much technical and mathematical detail. To balance out unwanted effects of working with partial solutions only in the other hour of exercises problems are posed and solved in small working groups without any additional written aids. All problems and corresponding partial solutions are given to the students at the beginning of the winter semester, solutions to the problems worked on during the small group exercises are handed out at the end of each exercise hour and all problems and solutions are available in an exercise brochure updated and printed annually.

A third brochure, which is also updated and printed annually, contains problems and solutions of the written examinations in the preceding three years. The basic idea behind the availability of the three brochures mentioned is to enable students who wish to do so to study all relevant subjects by themselves. On the other hand to all students a system of lectures and exercises is offered that will help them in obtaining the knowledge and proficiency deemed necessary in this and future times.

## 3. MODERN METHODS

Visual aids are employed to give a feeling for and some understanding of problems associated with dynamics of systems and processes leading for example to instability of ill-designed control systems. An example consisting of three cascaded tanks and a simple mechanical control of the level of the third tank has been realised on a PC and is shown by video projection. The same system is used also to show how frequency response functions for describing dynamic systems can be obtained from the measurement of sinusoidal inputs and corresponding output signals.

## 4. CONCLUSION

As control technology is more and more incorporated into systems and devices it has become an almost „hidden technology". Therefore the presentation of hardware-controllers is nowadays of much lesser importance as it has been one or two decades ago. For the same reason a well-founded understanding of what is happening in controlled and other dynamical systems is of outmost importance for the success of today's engineering efforts.

*Heinrich Rake*
*Aachen University of Technology*
*Aachen, Germany*

M. MUKUNDA RAO

# BIOMEDICAL ENGINEERING EDUCATION: A CASE STUDY

**Abstract.** Biomedical Engineering Education is a bridge between medical and engineering faculties and the student who is trained by both of them is expected to enhance the capabilities of medical services to society at large. Biomedical engineering education is imparted to the aspiring students both at the undergraduate (UG) as well as postgraduate (PG) levels. This interdisciplinary teaching program is increasingly becoming popular worldwide and the cream of student community is attracted to it sometimes preferring it to even Information Technology programs. In this paper, a case study of a PG program put together by two leading medical (SRMC & RI) and engineering (VEC) schools for a one year course leading to P.G. Diploma in Biomedical Engineering is presented and discussed. It is expected that it will serve as a link between full-fledged UG and PG programs in this area which are already in vogue.

## 1. INTRODUCTION

Gesundheit ist nicht Alles,
Aber, Ohne Gesundheit,
Alles ist Nichts !

--- A German Proverb

Health is not everything,
But, without Health,
Everything else is nothing !

--- English Translation

As per the above saying, nobody can dispute the importance of health in the well being of any society. Biomedical Engineering is an interdisciplinary area, whose objective is to apply effectively the engineering, science and technology to devise practical solutions to medical problems. This interdisciplinary area is aimed to meet the diagnostic, therapeutic and research aspects towards the well being of the humanity. In a broad sense, it embraces all aspects of science & technology to living systems. The justifications for starting such a teaching program in this area in a developing country like India are:
1. Compared to other developed countries, we in India are very much behind in this field even though our population has recently crossed one billion.
2. Requirement of trained personnel for our country in this area is very much needed to bridge the gap between clinician, engineer & industry, mainly due to the extraordinary developments in diagnostic and therapeutic equipment in recent years.

In view of this, two premier educational institutions in the area of medicine - SRMC & RI (D. U.) at Chennai and in the area of engineering – VEC at Vellore, which are separated by approximately 140 km but well connected by road & rail, have joined hands to design such an educational program. The main objective of such a program is to impart formal biomedical engineering education & training to the engineering and medical professionals. Besides, it is also intended to equip them with the capability for small/medium scale R&D for the benefit of clinicians and bio-medical equipment industry.

## 2. PROPOSED REGULATIONS

(a) Eligibility: Candidates, who are qualified graduates of engineering in Electronics & Communication, Electronics & Instrumentation and M. Sc. (Electronics) are eligible for admission. M.B.B.S. graduates who have an aptitude for engineering sciences and instrumentation are eligible for admission. Graduates from science specialties are not eligible.

(b) Duration & Commencement: The duration of the program shall be 12 months broken into 4 blocks of 3 months duration each. The course will ordinarily commence on $1^{st}$. August of the academic year.

(c) Medium of Instruction: Since India is a multi-lingual state with as many as 14 official languages, English shall be the medium of instruction for the subjects of study and for examination purposes.

(d) Contact Days & Attendance: Each academic year shall consist of not less than 180 contact days. Every candidate is required to put in a minimum of 75% attendance, both for the theory and practice/laboratory, in order to qualify him for the first appearance to the examinations.

## 3. SUBJECTS OF STUDY

(a) Block – I ( At SRMC & RI ):
    1. Medical Physics
    2. Human Anatomy
    3. Introduction to Biomedical Engg.

(b) Block – II ( At VEC ):
    1. Bio-medical Instrumentation
    2. Bio-medical Equipment
    3. Microprocessor based Medical Instrumentation
    4. Bio-signal processing

(c) Block – III ( At VEC / SRMC&RI ):
    1. Radiology & Imaging Systems
    2. Medical Information & Networking
    3. Medical Software
    4. Elective:
        (i) Lasers in Medicine
        (ii) Digital Image Processing

(d) Block – IV (At VEC / SRMC&RI / Industry):
   1. Seminars
   2. Industrial/Institutional Training
   3. Individual Project

## 4. DISCUSSION

The need for a suitable biomedical engineering education program in a developing country like India, whose population is over one billion and counting, cannot be overemphasized. For an effective health delivery, especially among the rural areas where the majority of the Indian population lives, the teamwork between medical & engineering disciplines is essential. It is in this spirit the present project is conceived and proposed. The days of confrontation between medical & non-medical fraternities, as epitomized by the following incidence cited below, are mercifully over.

> *"Will you allow me to ask you one question, Mr. Boon" said Andrew "Have you heard of Louis Pasteur?"*
>
> *"Yes", Boon was startled into a reply, "Who hasn't?"*
>
> *"Exactly! Who hasn't? You are probably unaware of the fact, Mr. Boon, but perhaps you will allow me to tell you that Louis Pasteur, the greatest figure of all in scientific medicine, was not a doctor. Nor was Ehrlich – the man who gave medicine the best and most specific remedy in its entire history. Nor was Haffkine – who fought the plague in India better than any qualified gentleman Doctor has ever done. Nor was Metchnikoff, inferior only to Pasteur in his greatness"*
>
> From the Novel 'The Citadel' by Dr. A. J. Cronin[1]

## 5. CONCLUSION

GOD is too important to be left alone to the <u>priests</u>

COUNTRY is too important to be left alone to the <u>politicians</u>

WAR is too important to be left alone to the <u>generals</u>

HEALTH is too important to be left alone to the <u>doctors</u>

---

[1] Where the hero of the Novel Dr. Andrew was sought to be impeached by the Medical Council of England (Represented by Dr. Boon) for joining hands with a non-medical professional in treating a patient.

## 6. ACKNOWLEDGEMENTS

The author is deeply indebted to Dr. T. K. Parthasarathy, Vice-chancellor of SRMC & RI and Mr. G. Viswanathan, Chairman of Vellore Engg. College for their encouragement & support to this program. My colleagues both at SRMC & RI and VEC have greatly contributed in designing this program. Financial support from the Alexander von Humboldt foundation of Germany towards the travel for participating in the prestigious $3^{rd}$ International Workshop on Global Engineering Education (GEE'3) held at the Technical University of Aachen / Germany during October 2000. Thanks are due to the organizers of this workshop for their kind invitation to present this paper at this prestigious workshop and for waiving the registration fees.

*M. Mukunda Rao*
*Sri Ramachandra Medical College & Research Institute (Deemed University)*
*Biomedical Sciences Division*
*Porur, Chennai, India*

T. PFEIFER, L. SOMMERHÄUSER

# QUALITY MANAGEMENT

*How to Manage Education and Research in a Research Institute*

**Abstract.** Quality management is already a successful concept for industrial companies. First experiences with a systematic quality management in research institutes and chairs have been gathered at the Laboratory for Machine Tools and Production Engineering (WZL). The article shows where and how a systematic quality management can be used to improve the performance of institutes and chairs in education and research. Therefore in a first step the customers, products and processes of institutes and chairs are analyzed to define areas where systematic quality management can help. In a second step the ideas and approaches of a systematic quality management are described. Some of these approaches are exemplified by concrete solutions, which are developed at the WZL.

## 1. INTRODUCTION

The situation of universities has changed in the last decades. Coming from a seller's market, where there were more students who wanted to study engineering than opportunities for prospective students, nowadays we find a consumer's market, which is characterized by a competition between universities for the best of the decreasing number of students (Figure 1).

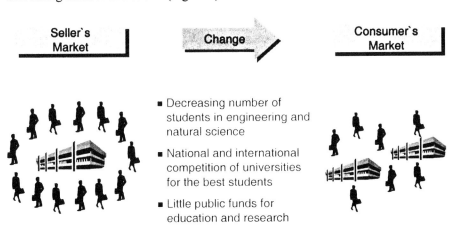

Figure 1. Changes in the situation of universities in Germany.

## 2. CUSTOMERS, PRODUCTS AND PROCESSES

This increasing competition and other factors force universities to improve their performance. To understand what is meant by performance in detail, it is necessary to describe the structure and functions of universities. These descriptions are valid for universities in Germany. The structures and functions differ from country to country. German universities are organized in three hierarchical levels (Figure 2). The organizational units on these three levels have different functions.

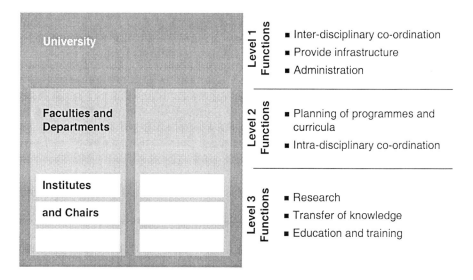

*Figure 2. Structure and functions of universities in Germany.*

The university as a whole has to plan and realize the co-operation between different disciplines represented by faculties and departments. Other functions are to provide an adequate infrastructure and the administration in the university. On the second level the functions of faculties and departments are the planning and controlling of programs and curricula as well as the intra-disciplinary co-ordination. On the third level the institutes and chairs have to carry out research work and knowledge transfer as well as education and training. This functional division in universities result in a special need for an effective and efficient management of education and research in institutes and chairs.

To improve the performance in education and research it is necessary to analyze the customers, products and processes of institutes and chairs. Therefore in a first step institutes and chairs must answer the question: Who are our customers and what do they expect from us? The different groups of customers must be identified. In universities there are external customers, e.g. students, sponsoring public organizations or companies, and in addition several internal customers, e.g. employees, faculties or other departments.

Merely listing the customers is not sufficient for customer orientation. Additional information about the expectations and need of these different groups of customers is necessary. Accordingly the products must be defined and the quality of these products must be characterized through the requirements of the customers on these products.

Products in universities are usually not material products, but services in the field of education or research. A practice-oriented education belongs to these "products" for the customer group students. Research result, which can be realized in profitable products or solutions, are examples for the customer group companies. As the products of institutes and chairs are primarily services the process "producing the service" cannot be handled separately from the product "service" [2]. The customer is highly integrated in the process of producing the service. In some cases the customer himself is the product of the service process.

An example: A student is the customer of an institute or chair. The coaching and supervision of his final thesis is a service which is given to him. This coaching and supervision is connected to the writing of the final thesis by the student. Thus, the processes in the institute or chair are closely related to the processes carried out by the customer. The goal of the coaching and supervision is to impart competencies to the student, which he can use in his coming professional career. Thereby the student is not only the customer of the university, but also its product. The customers for the product "educated and qualified student" are future employers like companies. Hence the design and management of the service "coaching and supervision of final thesis" must also meet the requirements of these customers.

The example shows that customers, products and processes are linked to each other. If the groups of customers are identified, then it is not difficult to define processes which lead to customer specific products and which produce a real benefit for these customers (Figure 3).

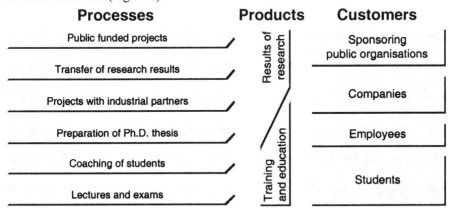

*Figure 3. Customers, products and processes of research institutes.*

The customers have special requirements concerning the services of institutes and chairs. These requirements have to be known before the service processes start.

Market analyses, interviews or surveys can help to define these requirements. In addition to that quality management methods like Quality Function Deployment (QFD) can help institutes and chairs to meet the defined requirements.

## 3. QUALITY MANAGEMENT SYSTEMS IN RESEARCH INSTITUTES

The next question is, how these processes and quality related activities can be designed, documented and managed. The suitable tool for this is a quality management system. It is the overall system to manage the quality and performance of an organization. In the quality management system the policy, objectives, customers, products and processes of the organization are defined. Concrete measure are developed and realized to support people doing their daily tasks and to improve their performance. Such measures are:
- Working procedures, which give people a clear understanding of the processes and tasks.
- A set of objectives, which provide people a guideline about what to achieve in the future.
- Support tools, like checklists, templates etc., which help people doing their daily tasks.

## 4. QUALITY MANUAL

The quality management system is described in the quality manual, which can be realized as an electronic documentation based on the hypertext technology (Figure 4). The benefits of the quality manual are:
- Reconsidering of existing processes
- Clear definition of processes and responsibilities
- High transparency of all processes and their interdependence
- Easy to demonstrate organizational capabilities to external partners
- Basis of an organizational Knowledge Management

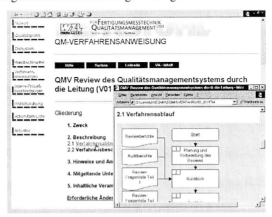

*Figure 4. Electronic documentation of the quality management system.*

## 5. QUALITY CONTROL LOOPS

Besides the customer information, which has to be investigated before the service processes start, information about the customer satisfaction has to be collected when the services have been produced. This information is needed to identify areas for improvement. The suitable tool for this are quality control loops.

Information itself has no value. Only when information is used specifically, it becomes an important success factor. This is also valid for customer information. Information about customer satisfaction must be used to improve the performance of the customer related processes. Therefore it must not only be defined how the information is collected. In addition to that it must be defined what will happen to the collected information. This leads to the definition of quality control loops [2].

The following quality control loop is set up to use the information about the customer satisfaction (Figure 5). Input for the process are the 5 M: man, machine, material, method and management. The processes are the controlled system. The output of the processes are services, e.g. research results. The service quality can be determined by questionnaires or interviews. The instrument customer survey can help to detect where the institutes and chairs failed to meet the customer requirements and to satisfy their customers. The institutes and chairs must define areas for improvement as well as suitable measures for improvements. The aim is to prevent deviation and customer dissatisfaction in the future.

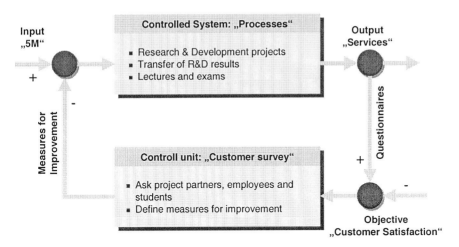

5M: man, method, material, machine and management

*Figure 5. Quality control loops in research institutes and chairs.*

An important support tool to measure the customer satisfaction are customer specific questionnaires, which are related to the services offered by the institutes and chairs [3]. Questionnaires can be developed e.g. for students, project partners or participants of workshops and seminars.

## 6. CONCLUSION

Many people and organizations have to work together to realize an excellent engineering education. It is the assignment of research institutes and chairs to educate, train and coach the young engineering students. Our experiences from the past three years have shown that a systematic quality management is of benefit for the performance of institutes and chairs in the fields of education and research. The employees and the whole organization are aware about the customer and his needs. People are higher motivated to produce a benefit for the customer and to secure the customer satisfaction. The awareness of the meaning of quality and the higher customer orientation itself increase the effectiveness and lead to performance improvement in the institutes and chairs. The customers have the opportunity to voice their expectations and needs as well as their satisfaction. By this the customer relationship and loyalty to the institutes and chairs could be improved.

Beside these implicit effects many concrete improvement measures have been developed and realized. To sum it up it can be said that introducing a systematic quality management in a research institute or chair require a lot of effort, but the benefits, institutes and chairs can derive from a systematic quality management, justify the effort in any case.

## 7. REFERENCES

[1] CORSTEN, H.: Betriebswirtschaftslehre der Dienstleistungsunternehmungen: Einführung, 2. Ed., München, Wien: Oldenbourg Verlag, 1990
[2] PFEIFER, T.: Qualitätsmanagement: Strategien, Methoden, Techniken, 2. Ed., München, Wien: Carl Hanser Verlag, 1996
[3] PFEIFER, T.; WUNDERLICH, M.: „Establishing quality systems in research institutes: a progress report", in: The TQM Magazine, 9 (1997) 3, p. 221-227

*Tilo Pfeifer, Lars Sommerhäuser*
*Aachen University of Technology*
*Aachen, Germany*

D. BRANDON

# ATTRACTING THE NEXT GENERATION OF STUDENTS

**Abstract.** The activities of the Technion Center for Pre-University Education include both 'second-chance' matriculation courses for candidates seeking entry to undergraduate programs, as well as enrichment courses in science and technology for children of all ages and several 'support for excellence' programs targeted at children with exceptional mathematical or scientific skills. The cultural and socio-economic diversity of Israel presents some unusual challenges that can only be overcome by close cooperation between the school system and the university. In particular, there is clear evidence that large numbers of children do not have the opportunity to develop their latent talents into the academic achievements needed to assure them of a place at university. Providing the educational opportunity, sustaining the student's motivation, and assuring that achievement is rewarded are tasks that we have to fulfill.

## 1. INTRODUCTION

Each society should have the freedom to decide its own priorities, each culture should preserve its own heritage, and each nation has to develop its own self-image. Globalization has its limitations and the well-publicized (and globally-organized!) opposition to globalization has to be taken seriously. At the same time, the benefits of the free transmission of goods, services and information across national borders are easily demonstrated and generally accepted, and the ever increasing demand for professional engineering expertise transcends cultural barriers and national frontiers.

Israel is a tiny country at the Eastern end of the Mediterranean that gets far more than its fair share of column-inches and sound bytes in the media. The Technion – Israel Institute of Technology is the premier training ground for engineering expertise in Israel, but only one among many alternatives for professional training available in Israeli universities, technical training colleges and extension courses offered by branches of foreign universities in Israeli. The Technion undergraduate population is culturally heterogeneous. Although predominantly Jewish, a variety of cultural and religious traditions are represented: liberal-secular, Sephardi and Ashkenazi, orthodox and conservative; Western or Eastern European, North African, Russian and Ethiopian. The non-Jewish student population is predominantly Arab but includes Christians and Muslims, as well as Druze, Circassians and Bedouin, none of whom would normally think of themselves as 'Arab'. The cultural disparity between orthodox, Muslim girls and their secular counterparts from the former Soviet Union can barely be imagined, while the Technion also hosts several Palestinian students from East Jerusalem. Approximately 70% of the student population is male, but the proportion of male students in each specialty fluctuates wildly, from over 90% in Mechanical Engineering to less than 30% in Biology and Architecture.

The Technion Center for Pre-University Education was created two years ago by combining the Department of Pre-University Studies (university entry programs) with a number of previously independent youth enrichment programs and some experimental programs that provided support for exceptionally talented children. Today the Center includes all three types of activity with a single, unified goal, namely to provide the Technion with a steady supply of highly motivated and well-qualified candidates for all our undergraduate degree programs. These programs include not only the traditional engineering disciplines, but also most of the natural and life sciences, as well as computer sciences, architecture, town planning and business management, as well as rapidly evolving, interdisciplinary programs in biotechnology and information systems. In this present contribution we outline the rationale behind the activities of our Center for Pre-University Education and describe some of our current problems, in the hope that our limited experience may yet prove relevant for other universities who are also attempting to fulfill the needs of some very different technological societies.

## 2. CURRENT ACTIVITIES

Our university entry programs constitute some 60% of our annual budget and run throughout the year. We are licensed by the Israeli Ministry of Education to matriculate students in Mathematics, Physics and English at the highest, 5-point level. We also teach courses in Technical Hebrew, as well as Hebrew Language (for new immigrants). Many of our candidates come from disadvantaged backgrounds, seeking a 'second chance, after having left school without matriculating at the level required for university entry. Many have completed army service, and come to us only after an extended, three-year break in their studies. Many of our candidates are new immigrants, faced not only with completing matriculation in Mathematics and Physics, but also with the need to achieve fluency in Hebrew and competence in English. Approximately 400 of our students qualify each year for entry to one or other of the Technion's undergraduate degree programs (a success rate of just over 50%). Our other graduates are either accepted by other Israeli universities or go on to study at one of the country's technical colleges. Within the same framework of Pre-University Studies we provide refresher courses in Physics and Mathematics for candidates who have been accepted for undergraduate degree courses, but who matriculated 3 or even 5 years previously. We also teach some basic courses within the undergraduate framework: a course in Physics for students who lack the 5 credit points in physics required by the Technion, and a course in Graphics for freshman students of Architecture and Town Planning.

Our youth enrichment programs start in primary school (as early as grades 4 and 5, ages 9 and 10) and run right the way through junior and senior high school (to grade 12). They are distinguished by the absence of examinations and entry requirements, and any child suitably motivated may take part. Topics covered include any and all of the sciences and mathematics, as well as subjects intended to introduce children to activities associated with a professional careers in science or engineering. As would be expected, popular courses are those given in computer languages, the internet and robotics, but zoology and veterinary medicine are also very much in demand. All our enrichment courses are given weekly, in 25 two-hour sessions throughout the school year, and the same courses are available (in a truncated form) as a day camp held during the summer break. Schools can also attend class 'study days' on campus, visit various academic departments by appointment, and request visits to the school by a 'mobile laboratory' (where demonstrations of selected science topics are given at the school by our staff). The number of children registering for our enrichment programs each year is now in the thousands, limited primarily by geographical accessibility and the availability of public transport.

Our 'support for excellence' programs are currently our fastest growing activity. While the Technion has always found ways to allow exceptional students to begin their university studies early, it is only recently that we have actively sought to encourage these candidates in cooperation with their schools and the educational authorities. Three programs are now well established. Our TeLeM (Technion Lessons in Mathematics) program begins in the final year of primary school (grade 6) with mathematics enrichment courses for suitably motivated children. In junior high school (grades 7 to 9) talented children are encouraged to join a TeLeM class dedicated to completing the 5 credit point mathematics matriculation requirement. Exceptionally gifted children may join a 'fast-track' Technion Class and complete their matriculation requirements at the end of grade 10. Both the TeLeM and the Technion classes are given 8 contact hours in mathematics each week (in place of the 5 or 6 hours which is standard practice in Israel) and an additional one or two hours of afternoon enrichment.

Our second 'Excellence' program is one of Advanced Placement, in which teenagers with the necessary mathematics qualifications are allowed to enroll for undergraduate courses without having to register for a specific degree program. Credit points accumulated from studies successfully completed can be 'saved' towards a future degree program. Most recently, graduates of the TeLeM program have been joining the Advanced Placement program in sufficient numbers to justify providing teaching assistants at the school to help these students with their studies. Even more encouraging, the best students from the Advanced Placement program have now accumulated sufficient credit points to be accepted for direct transfer to undergraduate degree programs as full-time undergraduate students. The grade point average (GPA) of our Advanced Placement students (most of whom are in the 15 to 17 year old age group) is in the top 20% of regular Technion students.

Finally, the longest running 'excellence' program is the SciTech International Summer Research Camp, now in its 10$^{th}$ year. This camp is attended by of the order of 80 teenagers from all over the world who join us for a month of dedicated project work in one or other of the Technion's laboratory facilities under the supervision of our research and academic staff. A SciTech project has a beginning, a middle and an end. Successful candidates are placed in e-mail contact with their project mentors well before the camp starts and are required to prepare themselves for their projects by studying recommended texts. Since the students arrive fully prepared, no time is lost in starting work on the projects. In the last week of SciTech students are required to prepare project reports, as well as seminars and posters. Seminar presentations are made at a SciTech Workshop, which usually has three parallel sessions covering Mathematics and the Exact Sciences, the Life Sciences, and Engineering Science and Technology. At the closing ceremony prizes for excellence are awarded and students are presented with a certificate of participation and a copy of the annual SciTech Proceedings.

*2.1. Some Principles*

In all our activities the Technion Center for Pre-University Education is only one partner in what ought to be a joint effort involving the student's social environment, as well as his teachers and the educational authorities. The twin principles that have governed the development of all our programs are quite simple:
1. The single objective of every program has to be the conversion of latent talent into academic achievement through the presentation of achievable goals.
2. No Technion pre-university program can succeed without the active support and understanding of the student's family and friends as well as his teachers and the educational authorities.

Implementing these principles may be less straightforward. Turning talent into achievement requires an appropriate learning environment to attract the student to the program un the first place, on-going activities over an extended time in order to maintain motivation, and rewards and recognition for the student's achievements while completing the program. Similarly, 'support and understanding' for the student is necessary, not just when he begins a program, but throughout its duration and at its completion. Talent takes many forms, and cognitive talent in mathematics and linguistics has to be nurtured alongside talents for creativity, innovation, leadership, cultural sensitivity and communication. The teaching staff is, of course, critical, and an appropriate forum has to be found for teachers to exchange views and expand their teaching activities and skills. In our TeLeM program we have developed our own teaching materials and involved the teachers in the schools in the development and evaluation of these programs. In fact, we have refused to enter a school unless the mathematics teachers are prepared to accept this commitment.

*2.2. Some Problems*

There are some very general requirements for success in the engineering profession. A good engineer must be able to analyze a problem and synthesize a solution. He analysis and his proposed solution must be accurately quantitative and he must have the linguistic and communication skills to present his results to diverse audiences: his employers and the workforce, as well as his company's customers, competitors and suppliers. Classical engineering education commonly focuses on numerical analysis, leaving the student to develop communication skills outside the formal course framework, and with synthesis usually relegated to work on a final-year project.

In order to compete successfully in the learning environment at the Technion, students must have the capacity to absorb new knowledge rapidly and efficiently, and they need to be able to retain and apply this knowledge. Since acceptance to the Technion's degree programs is dominated by the student's matriculation grades, which are related to the extent of his knowledge at the time of graduation, rather than his ability to absorb and retain new knowledge, it is not surprising that many undergraduate students lack the learning skills needed to ensure good grades. At the Center for Pre-University Education the clearest illustration of this problem is the absence of correlation between initial test scores in either mathematics or psychometric tests and final entry grades to the undergraduate programs. This is illustrated in the figure, which shows the student's university entry grade plotted against his initial psychometric test score for two pre-university studies intakes (January and July, 1999). While a high psychometric test score (over 650) is a fair indicator of a student's ability to achieve the grades necessary for university entry, a large proportion of the candidates with poor test scores (30% of those scoring less than 650) are ultimately able to gain entry to a degree program.

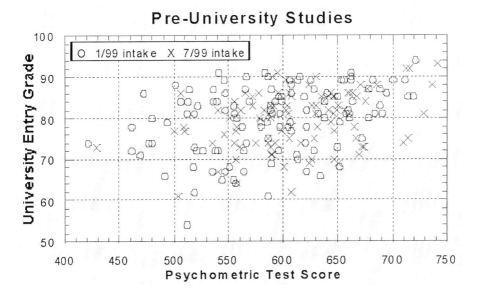

Most of these students come to us after completing their army service, at the age of 22 or even older, but similar problems exist in another program, initiated last year fro teenagers (aged 15 and 16) of Ethiopian origin who are seeking to improve their mathematics matriculation performance. The initial group of 64 teenagers attended a 4 day evaluation camp at which they were tested on their understanding of a series of graded lessons given on topics unconnected with their matriculation studies. This estimate of their capacity to understand and apply new knowledge was used to assign them to 4 different enrichment classes. The program proved to be a major success, with demonstrated improvements in school grades, student's self-image and the support of the families. A further group of 45 students has now joined the first group for the second year of the project.

Similar problems exist in evaluating candidates for the TeLeM program, and again the use of a graded set of mathematics lessons to evaluate academic potential has proved to be a useful tool. Approximately 10 hours of lessons and exercises seem to be the minimum required to develop a reliable picture. The lessons are set at three levels: first 'easy', then 'difficult' and finally 'achievable'. Borderline cases are followed up with personal interviews before a decision is made to allocate a candidate to one or other of the special mathematics classes.

We have not yet succeeded in implementing a similar program for the programs aimed at mature, 'second chance' students, who probably require more preparation to make this type of testing acceptable.

## 3. DISCUSSION

Attracting students to an engineering career requires that we make full use of the windows of opportunity presented by the natural development of the individual from child to teenager to young adult. We are missing the window of opportunity if negative symptoms of boredom, laziness, rebellion and alienation are ignored, and these symptoms should be pre-empted through the challenge of achievable academic goals, with sustained performance over an extended period of time, and by ensuring recognition for achievement from family, teachers and friends.

We believe that our university-based programs have to be fully coordinated with educational authorities as well as with the families of our students if we are to be successful in developing natural talent into academic performance. Too many potentially good candidates for the engineering profession fail to apply because they underestimate their own potential, or because they do not yet meet the formal entry requirements, and many gifted teenagers and young adults still lack adequate opportunities to develop their talents.

Student awareness and motivation need to be coupled with a reliable assessment of the student's learning skills, and nurtured within a supportive learning environment. Prejudice and discrimination are still common, not least in determining the expectations of students and their families, and need to be countered through cultural sensitivity and awareness. Our limited experience would seem to indicate that success is directly related to our ability to work with the schools, the teachers, the educational authorities and the student's families. The key is the recognition of a common goal.

## 4. ACKNOWLEDGEMENTS

The programs described in this contribution and the philosophy behind them have developed over an extended period of time and with the active involvement of a large number of individuals. The success is theirs, the misconceptions are my own.

Prof. Avi Berman (Faculty of Mathematics, Technion) first introduced me to the SciTech program and its originator, Harry Stern of New York. Avi was also the instigator of the Technion Classes in Mathematics and the Advanced Placement program at Technion (at least in its present form).

Dr. Orit Zaslavski (Department of Education in Science & Technology, Technion) was a prime mover behind the development of the TeLeM concept, whose implementation has depended heavily on the expertise and unstinting hard work of Dr. Roza Leikin (University of Haifa).

Dr. Dalia Pisman (Jewish Agency) was the prime mover behind the mathematics program for teenagers of Ethiopian origin. Her enthusiasm and guiding hand played a major role in turning this experimental effort into a major pedagogic success.

Many, many other individuals have provided support, information, inspiration or all of these. Prof. Yoav Benyamini, the Dean of Undergraduate Studies at Technion (my immediate superior) and Sara Zuker, the Secretary of the Center for Pre-University Education (my ever present right-hand) both deserve a special mention.

*David Brandon*
*Technion – Israel Institute of Technology*
*Center for Pre-University Education*
*Haifa, Israel*

# PART VII.

# SUCCESSFUL PRACTICE IN ENGINEERING EDUCATION -
## EDUCATIONAL CONCEPTS

J. DEPUYDT

# LEARNING INVENTIVE THINKING IN THE FORMATION OF ENGINEER-ARCHITECTS

**Abstract.** This paper proposes a conceptual framework in order to investigate the position of 'Theory of Architecture and /or Engineering' in a constructivist-based learning environment, as opposite to instructionist learning environments.
A described taxonomy can be approached in different contexts: the context of the course 'Theory of Architecture/theory of Engineering', the context of the 'Design Studio', the context of the curriculum of the formation, the context of the school or university as learning environment, the context of society and its professional attitudes...
The process of learning seems to be a process of invention, in terms of a balancing 'in-between' intention and intervention. It implies both conceptualization and technique. Inventiveness can methodologically be deployed in order to teach students to create their own philosophies and narratives, in terms of self-education.

## 1. INTRODUCTION

In the curriculum educating Civil Engineer-Architects at the 'Free University of Brussels', some aspects attract attention by comparing its curriculum with the other 'Engineering' courses.

Firstly, at the department Architecture, a course entitled 'Theory of Architecture, seen as the philosophical core of the discipline exists, whereas at the departments of Engineering, there is(are) no course(s) entitled 'Theory of Engineering'; however all sub-disciplinary courses contain a theoretical part, e.g. 'Theory of Mechanics' as complementary to a practice in terms of exercises, practice, laboratory work... By 'Theory of Engineering', I mean the integrative structure of engineering concepts, including historical, cultural, technical, ...aspects.

Secondly, the course 'Architectural Designing', which occupies an important place in the Design studio, is a typical component of the Civil Engineer-Architects curriculum. The design studio can be defined as a mental place of dialogue, where all sorts of knowledge (scientific, technological, humanistic...), skills and attitudes are integrated.

Two arguments can put foreword here:
a) What is the real sense dividing a curriculum into sub-disciplinary-based courses, regarding the fact that one of the main objectives at the end of the education is an integrative and constructivist-based understanding and insight in the discipline (its knowledge, critical reflection and production of new knowledge)? The discipline involved is 'per definition' multidisciplinary: Engineering and Architecture.
b) What is the real sense detaching theory from practice, regarding the fact that in a constructivist-based learning environment (1), there is no course in the curriculum entitled 'Theory of Architecture'? On the other hand, there's a

permanent awareness of the theoretical implications of the architectural design practice in the Design-Studio, which is a learning environment for the reflective practitioner (2).

These two arguments tend to recognize that a constructivist learning environment based on design projects/problems, where learners built up a sustainable, dynamic and conceptual (3) knowledge structure is a very important both professional and intellectual 'environment'.

## 2. CONCEPTUALIZATION AND INTEGRATION

Increasingly, there is a demand of active -learning environments in the formation of Engineers in general; based on thematically blocks, problems, and projects... All of these are aiming to 'integrate' knowledge, skills and attitudes (4).

On the one hand, in most cases, there is no global 'theoretical' framework for what 'Engineering ' is, and what is taught in engineering disciplines. However it exists under the form of 'philosophy of science.

On the other hand 'Theory of Engineering' referring to the theory and praxis of 'Technè', technique, technology, tectonics...structure, construction) is apart of 'Theory of Architecture' whether the last one is defined as 'Philosophy of architecture' or 'Design Theory'. Because of that, an insight in some 'exportable' pedagogical aspects about the formation of engineer-architects could be very productive.

Since there is a strategy in defining the particularity of what (research in) Engineering-Architecture is instead of doing teaching and research in separately architectural and engineering matters, the formation of Engineer-Architects has the benefits:

a) Having a course 'Architectural Design' (the Design Studio) as part of the curriculum; where in potential all sorts of knowledge, skills and attitudes are integrated.
b) Having a course 'Theory of Architecture', delivering a conceptual framework/language in order to build up and dealing with the knowledge-body constructed by students themselves. The pedagogical principle of student-centered learning can be dominant and productive in this respect.

The main questions then are:
a) How can we exactly take advantage of these benefits?
b) How can we exactly communicate these benefits, in order to export them to other disciplines?

Generally an analogy between the act of 'knowing', 'investigating' and 'designing' underlines the constructivist approach. The design studio seen as a place where knowledge can be shared, experienced, applied, communicated and evaluated is the place where knowledge is conceptualized and applied simultaneously. This is a basic integration process in action.

Based on literary study, subsequent models are built up in order to analyze the position of the 'Design Studio' (content, context, skills and attitudes), the position of 'Theory of Architecture' (content, context, skills and attitudes) and the relationship between those in existing architectural education. Key concepts like active learning, interdisciplinary; dialogical learning, diagrammatically discourse, conceptual integration and invention ('ars inveniendi') are ruling phenomena. The last one, the inventive skill, takes a central place, since it can be understood as that sort of creativity which enables, by finding new connections between existing constructs, to determine new constructs (inventions). A sub-model (Table 1) is framing following elements:

*Table 1. Elements of a conceptual framework on 4 axes*

| Axis A | Learning | Cognitive | Behaviorist | Constructivist |
| Axis B | Instructing | Knowledge | Reflection | Research |
| Axis C | Doing | Technique | Design | Culture |
| Axis D | Thinking | Procedure | Creativity | Invention |

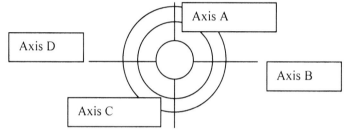

This matrix can be used in differentiating topics and levels in the courses e.g. 'Theory of Architecture'.

In seconds year e.g. 'Theory of architecture' can be knowledge-based, respectively thirds year reflection-based, respectively fourth year research-based (5). Encouraging cross-relations between Technique, Culture and Design in acting as an Engineer-Architect at all levels, broadens the existing internal vitality of Theory of Architecture driven by architectural history, towards a dynamic understanding about the role of design and technology and contemporary culture (6).

Referring to Moholy Nagy's model (Table 2) in the Bauhaus pedagogy, plastic elements are related to tools and media (technological dimension) and tools and media are related to plastic elements (esthetic / theoretical dimension).

*Table 2. Architectural Tools/Media*

| Plastic Elements | | | Tool/Medium | | | Plastic Elements |
|---|---|---|---|---|---|---|
| ========== | | | ========== | | | ========== |
| Line   | * | * | Brush | * | * | Line   |
| Shape  | * |   | Light | * | * | Shape  |
| Color  | * |   | Paper | * | * | Color  |
| Space  | * |   | Wood  | * | * | Space  |
| Motion | * |   | Clay  | * | * | Motion |
| ....   | * |   | ....... | * | * | ....   |

It is important to emphasize that not only the differentiated field in the matrix is of any importance, but the connections between those in order to provoke a new conceptual understanding.

Like good Design is based on balances, a 'good' curriculum is based on a good balance in treating and experiencing the fields of above described model. Such a balance between 'Learning' and 'instructing' and between 'Doing' and 'Thinking' has to be 'invented' as well.

*Table 3. Model of contexts*

Above described matrix can be approached as a field, within different contexts: the context of the course 'Theory of Architecture/theory of Engineering', the context of the Design Studio, the context of the curriculum of the formation, the context of the school or university as learning environment, the context of society and its professional attitudes... (Table 3)

In constructivist learning environments, the 'construct' is the central component of the learning process. This process can be made visible, debatable and assessable in the 'design studio' (as place) during the course 'Architectural Design'; (sometimes the only component of the curriculum (1).

More than being a physical place, a real 'Design Studio' is a way of thinking, where complex functional and intellectual values are integrated, it's a place of dialogue, reflection on professional practice (*)(7). Otherwise, the Design studio is an 'empty' place.

All of these contexts can be seen as institutions (8), with their own context-dependent governing practices, conventions, experiences, values and language(s). The contents of these contexts are after all 'relative constructs'; since they can be transformed.

Important is the aim to connect different contexts. Therefore, the content of each context e.g. 'the curriculum' is less important than the relation between 'the design studio' and 'the curriculum' and the relation between 'the curriculum' and 'the school/or university'. Such a connection can be defined as a simultaneous treatment of rules in alternative contexts.

It makes no sense to interchange the values between the con-'texts' /but you can exchange them in terms of simulation. Designing is therefore trans-structuring by means of conceptualizing, ordering by means of putting into operation and doing interventions by means of téchne/the making.

The ability in mastering culture values in different con-texts is a main attitude for (architectural) designing. Constructing connections between different contexts activates building up 'experience' in terms of 'seeing, interpreting and doing'.

Metaphorically we can introduce here Friedrich Fröbels concept of the 'Kindergarten'; using outside domain of the school in order to activate real experience.

Relations between the different scaling levels (Theory of Architecture/ Engineering, Design Studio, Real Practice, Utopic practice, Theory of the discipline') are stimulated by a double attitude: intention and intervention- the way of 'perceiving' and the way of 'making'.

It is in the area of tension between 'intention' and 'intervention', that the concept 'INVENTION' is characterized, as an interplay between 'diagnosis' and 'prognosis'.

The hypothesis is that three vectors characterize an architectural pedagogy; the way of perceiving the real world (a context in general), the way of its interpretation/treatment and the way of communication and relation with the real world (a context in general): intention, invention, and intervention. Therefore the power of invention is a main concept in an experimental design studio and interwoven with the power of de-/trans-/re-structuring knowledge constructs.

Prof. Rob Cowdroy (9) defines 'research', since it is a mental activity, equal to 'design'. In both, he is making a difference between three levels:

A/ incremental/procedural, given/determined = PROCEDURAL ACTING/ applying rules.
B/ developmental/application, perceived/adherence to an established body of theory = low level CREATIVITY/ combining rules in order to create new things by combining already existing ones in response to a perceived need.
C/ fundamental/new theory, conceptual solution and breakthrough = INVENTION/ generating new rules.

Invention, which can be defined as high-level creativity, occurs at the third level, where a new theory arises resulting from multi- and interdisciplinary activity and experimentation. This principle challenges the focus on specialization.

The main question here is the following: are today's research-oriented engineering instructors able to help their students to become inventive engineers in their future workplace?

The proposed answer is YES:
- if there is a balance between profession, designing and research.
- if the 'real' experiences are implemented in the learning environment in order that a learner attempts to make sense of his/her experiences enlarging the already existent knowledge-construction and knowledge-structure.

INVENTION occurs when learners are liberated from old ideas, in order to encourage the process of changing concepts and perceptions. "Developing an environment which allows for the freedom, (...) to allow for creativity and then the motivation required to generate autonomy in students" (10).

Going back to the different contexts in which the 'theory of a discipline', the 'Design Studio', the curriculum, the culture of a school/university, society... are related; all of these contexts are 'institutions' in the sense that they are rule-driven. Designing the content of all of these contexts is determined by the larger context.

Generating the rules, as a social construct, is a process of 'giving meaning', which is an activity 'an sich'. An appropriate language is necessary in order to make cross-border communication possible. The 'institution' 'Theory of Architecture/ Theory of Engineering' is delivering that appropriate language for the discipline.

In a similar way, each field can be split up in a similar model e.g. (Table 4- Professional Practice).

Situated on the axis B/instructing and C/doing.

*Table 4. Professional practice*

| PROF. PRACTICE | TECHNIQUE | | | | CUL-TURE | DESIGN |
|---|---|---|---|---|---|---|
| KNOWLEDGE | Tech | Tech-nique | Culture | Design | | |
| | | Knowl. | | | | |
| | | Reflect. | | | | |
| | | Resear. | | (*) | | |
| REFLECTION | | | | | | |
| RESEARCH | | | | | | |

We can argue that there exists the interplay between:
A. 'Content-knowledge' and 'process-knowledge'.
B. 'Obtaining new solutions for a known problem' and 'the use of known solutions for a new problem' (11).

## 3. A LEARNING ENVIRONMENT FOR CONCEPTUALIZATION AND INTEGRATION

Both, seeing the 'problematic' and re-making the 'problematic' are creative acts which exceeds de procedural and systems-based design processes, and are based on an intuitive power. However, in 'connecting' both, real inventions can occur.

Three partial models support the general model:
a) Relations between knowledge-based courses and the design studio, and a proposed hierarchical structure in the curriculum construction.
b) Relations between the technological and cultural context through qualities and means/skills and materials (12).
c) Relations between the real world and the design studio. Prior to the concept of the individual creative and inventive genius is the concept of beings totally engaged in a real world.

Learners become constructors of personal products/constructs (works, oeuvres, design-products, case studies...papers, essays) which can be communicated with others. Further research is needed in order to map the sort of constructs that are used in a constructivist learning environment, or are the result of a constructivist learning process. This constructivist approach is applied on a study about the position of the 'Design Studio' and 'Theory of Architecture' in the curriculum and the place of experiment concerning the philosophy, ethics, aesthetics, design and practice of the 'construing' and 'constructing' of the world (13).

Since learners must participate and interact with the surrounding environment in order to invent and negotiate their own view of the subject and since designers think in models, diagrams and drawings, the integration of knowledge and skills and the integration of an academic/research context and professional context is evident.

Moreover, inventiveness can methodologically be deployed in order to teach students to create their own philosophies and narratives, in terms of self-education. Therefore the 'construction' of theories, empowering 'design-talk' is more important rather than 'theories'. Student-centered education can therefore be a condition.

## 4. THE 'INVENTIVE' POWER IN THE CENTER

The 'inventive' power in the center, between 'conceptual power' (axis C-D in the model) and 'dialogical power' (axis A-B in the model) on the one hand, between 'intention' (intellectual-academic approach of the problem) and 'intervention' (concrete-pragmatic approach of the problem) on the other hand, claims generating new concepts, by inter-relating aspects from the four axes (Learning, Instructing, Doing, Thinking) of the taxonomy in describing a sustainable learning environment.

Concepts of scale and context on the one hand, attention for hermeneutic, historical and critical knowledge on the other hand are characteristic of 'Architectural Design.' These can be exported to other Engineering disciplines; the formation of an 'Engineer-Architect ' is an implicit invitation.

The above described model admit to (re) structure architectural concepts and concepts of architectural education; looking for significant connections, or looking for particularities, intending integration or differentiation, on different levels without hierarchical distinction (Architect-Engineer or Engineer-Architect).

* Theory of architecture I- (seconds year) as learning to understand/perceive the 'real' world through the dynamic knowledge of architectural concepts, imbedded by the use of drawings/diagrams, photographs, model... as an appropriate skill.
* Theory of architecture II- (third year) as learning critical reflection on 'real' world', imbedded by the use of texts (writing and reading a paper, a personal autobiography, texts, critiques, manifestoes...) as an appropriate skill.
* Theory of architecture III- ($4^{th}$ year) as learning to produce new architectural knowledge through researching one or more aspects and/or through stimulating effective 'design talk' in the 'Design Studio'.

In order to strengthen the constructivist-based approach, the position of student activities and their products (papers, drawings, written reports, oral reports...) in the thirds year 'Theory of architecture' e.g. students were asked to write a 'scientific autobiography' (14), in which they have to theorize the design projects they have designed in the first and second year 'Design Studio'.

Experience as Design-Studio teacher has resulted in a pedagogical approach where students treat one complex and real problem during the whole year (15). In fact they are for several times re-designing the same problem. The first time is an introduction to the problem and its complexity. In order to initiate the students in the character of the problem, they have to find a 'design-solution' without specific knowledge. In an atmosphere of 'learning by doing, experiencing, playing...', the students increase their motivation and engagement with respect to the problem. After a period of some weeks, they already have a design; a tangible product by which they can launch and share experience. Each time they have to re-design the same problem, specific knowledge, skills and attitudes are explored in depth e.g. organizational-functional aspects, ink-drawing and pragmatic approach to a problem. On the other hand they have to integrate already acquired knowledge, skills and attitudes from the former re-designs. At the end of the year, during the final re-design all knowledge, skills and attitudes are integrated. In terms of feed-back, the students can compare the result with the first design, and 'see' their own learning process; as a permanent re-structuring and re-constructing process.

This can be productive in order to experience and to experiment with the complexity of design strategies (16). After all, the design studio is a 'research laboratory'.

In each year and through all years a balanced relationship between e.g. 'Technique', 'Culture' and 'Design' seems to be necessary.

From theoretical point of view, the construct "Problem-exploring in depth-Problem" can be related to Moholy Nagy's model, whereby an aspect of the problem (e.g. character) is connected with different Architectural tools and media (drawing, function, program, structure, form...), on the one hand (technological dimension), whereby one aspect of Architectural tools and media (e.g. structure) ) is connected with different aspects of the problem (ecology, situation, character, economics,...), on the other hand (esthetic/theoretical dimension), (Table 5).

*Table 5. Problems/Architectural tools and media*

| Problem | | | Architectural Tool/Medium | | | Problem |
|---|---|---|---|---|---|---|
| ========== | | | ========== | | | ========== |
| Ecology | * | * | Drawing | * | * | Ecology |
| Situation | * | * | Function | * | * | Situation |
| Humanities | * | * | Program | * | * | Humanities |
| Character | * | * | Structure | * | * | Character |
| Economics | * | * | Form | * | * | Economics |
| .... | * | * | ........ | * | * | .... |

The technological dimension can be related to INTERVENTION, such as the esthetic/theoretical dimension can be related to INTENTION.

INVENTION is considered in-between.

## 5. CONCLUSION

The process of learning is a process of invention, in terms of a balancing 'in-between' between intention and intervention, which implies conceptualization and technique.

Strategies in connecting e.g. 'Technique', 'Culture' and 'Design' on the one hand, 'Knowledge', 'Reflection' and 'Research' on the other hand, has to liberate 'Theory of Architecture' seen as a course, small part of the curriculum, from its separate position and connect its content to 'technical' and 'design'- matters. A constructivist-based learning environment is therefore desirable.

At first, theory is 'Design talk'.

This idea reinforced the idea of 'design activity' as a 'research activity' and is the rationale for 'inventive' thinking; something that can be taught and learned; keeping in mind that "learning is changing" (17).

Brought together with the principles of Life-long Learning, a ground for sustainable Design and Research is developed. (18) Learners have to become constructors of personal products/constructs (works, oeuvres, design-products, case studies...papers, essays) which can be shared with others.

Since learners must participate and interact with the surrounding environment in order to invent and negotiate their own view, the integration of knowledge and skills and the integration of an academic and professional context is evident. The power of invention is a main concept in an experimental design studio. Moreover, inventiveness can methodologically be deployed in order to teach students to create their own philosophies and narratives, in terms of self-education. Student-centered education is therefore a condition.

The aim of learning is to handle changes; therefore inventive thinking, defined as pushing out the boundaries of the before mentioned con-texts, is a long-lasting strategy.

## 6. ACKNOWLEDGEMENTS

Prof. Dr. Rob Cowdroy (University of Newcastle/Australia) and Prof. Dr. Gerard Van Zeijl (Technical University of Eindhoven).

## 7. NOTES AND REFERENCES

(1) such as 'The integrative Problem Based Learning –model' in Newcastle/Australia
(2) Referring to D. A.Schön, "Educating the reflective Practitioner", Jossey-Bass Publishers, San Francisco, 1987
(3) A concept is a mental representation of a category, which allows a person to sort stimuli into instances and noninstances.
(4) Referring to different conferences held by SEFI - Société pour la Formation des Ingénieurs.
(5) The second year course is leaning on the first year course 'History of Architecture' and the introduction in Architectural designing 'Design Methodology I'.
The 3rd-year course is preparing to the 'Research Elective' and 'Design Elective' in fifth year.
(6) John Dewey was defining technology as an art of experimental thinking.
(7) Prof. Rob Cowdroy- University of Newcastle/Australia.
(8) Referring to the 'Institutional pedagogy' of Fernand Oury.
(9) Prof. Rob Cowdroy (University of Newcastle/Australia.) ref. EAAE/ARRC conference Paris july 2000 - "Point zero".
(10) "Fostering Creative Thinking in student Engineers"- Caroline Baillie and Paul Walker, in 'EJEE, Volume 23 number 1, March 1998.
(11) R. Sell, "Beteiligungsqualifizierung." Aachener Reihe Mensch und Technik Bd11, Aachen, 1995.
(12) Referring to Moholy-Nagy's model for the methodological structure of the Preliminary Course in Chicago.
(13) Referring to M. Franscari : Marco Frascari, "The Tell-the-tale Detail", in Kate Nesbitt (ed), "Theorizing a new agenda for architecture-an anthology of architectural theory 1965-1995", Princeton Architectural Press, New York, 1996.
   - Constructing: the result of the logos of techné.
   - Construing: the result of the techné of logos.
(14) Inspired by Aldo Rossi's 'scientific autobiography'.
(15) Design Studio 1st year 2000/2001- Provide an accommodation for homeless people in Brussels by renewal of an industrial site.
(16) Prof. G. Van Zeijl – Evaluation of the course 'Theory of Architecture' at the Free University of Brussels within the scope of the visitation report. -July 2000.
(17) Ref. "To learn is to change. Education is a process to change the learner."
(18) Therefore, the different described partial tables can be combined. In the scope of this paper, this is not developed.

*José Depuydt*
*Free University of Brussels*
*Department TW/Architecture*
*Brussels, Belgium*

E. ESPOSITO, E. SIGLER

# ACQUIRING THE TOOLS TO BECOME A SUCCESSFUL ENGINEER IN THE 21$^{ST}$ CENTURY : APTITUDES AND ATTITUDES

**Abstract.** The major objective of Ecole Centrale Paris (ECP) has not changed since its foundation in 1829: training high-level engineers for industry. The majority of ECP alumni will become top-ranking technical specialists, entrepreneurs or managers.
Today, a modern engineer should be able to work on a team and to analyze and then solve complex problems in an international environment. So s/he should be a professional and an accomplished communicator, open to international realities, desirous of constant improvement, and finally both modest and with a moral ethic.
In such a context, social and human sciences have an important part in the curriculum (which covers three years following two years of intensive advanced mathematics and physics). It emphasizes a balance among three fundamental aspects: acquisition of a broad, multidisciplinary scientific and engineering background, development of interpersonal skills and problem-solving attitudes, and an introduction to the dynamics of the engineering profession underlining the importance of innovation, creativity and questioning capacities. The international aspect comes from a widely followed double-degree program (the TIME Association), and strong interaction with industry exists throughout the curriculum.

## 1. INTRODUCTION

Four years ago, after consulting alumni and other experts in industry, and following a two-year process of internal preparation, Ecole Centrale Paris launched a reform of its engineering curriculum to adapt it more closely to the needs of industry in the upcoming century. The conclusions of the experts were that if academic excellence was a given and knowledge of the basic engineering disciplines essential, these were not sufficient to guarantee students would become successful engineers. The school consequently initiated new learning and teaching strategies for its students.

This paper presents the French system of higher engineering education, the assumptions underlying the new strategies, a presentation of the revised curriculum, and a succinct description of the ongoing work.

## 2. THE FRENCH SYSTEM OF HIGHER ENGINEERING EDUCATION

### 2.1. The « Grandes Écoles »

The French educational system in engineering is comparable to those in the rest of Continental Europe, but quite different from those in English-speaking countries.

Engineering studies in Continental Europe produce two different profiles of engineering degrees: one with a short curriculum (three to four years long) based on an applied approach, and one with a long curriculum (five to six years long) based on a theoretical and conceptual approach. In contrast with the Bachelor/Master in which the shorter curriculum is commonly part of the longer one, the two curricula are different from the outset.

In France, the best students of an age class are attracted by engineering studies, and consequently the potential of the student audience differs from that of engineering departments in most other countries. Most engineering schools in France, called *Grandes Ecoles d'Ingénieurs*, are independent of universities and train top industrial and administrative managers.

*2.2. The « Classes préparatoires »*

The *Classes Préparatoires* correspond to two years of post high-school education with heavy emphasis on mathematics and physics. The first year of the *Classes Préparatoires* is termed *Mathematiques Supérieures* and corresponds to the freshman year at US universities. The second year is called *Mathematiques Spéciales* and corresponds to the sophomore year at US universities. At the end of the second year, students take the nation-wide competitive entrance examinations to what are called the *Grandes Ecoles d'ingénieurs*, including Ecole Centrale Paris. Thus, all students who enter Ecole Centrale Paris have at least two years of advanced-level post-high school education.

*2.3. The Process of Admission to Grandes Ecoles*

Admission to the engineering schools in France is a very selective process: of the 800,000 young people of an age class, approximately 400,000 pass the *Baccalauréat*, the French secondary school final examination, each year. The best students normally take the Scientific *Baccalauréat* in fundamental sciences, and of these 100,000 pass. Students who are awarded the *Baccalauréat* with honors can go on to the *Classes Préparatoires*.

After two years of university level study, 12,000 of them take the entrance examination to Ecole Centrale Paris. ECP admits 330 students a year from among the top 700 science students in France. More than 60 additional students from scientific universities, mainly in several European countries, follow the first two years of studies on an accredited double degree transfer exchange scheme (the TIME Program).

## 3. THE ASSUMPTIONS UNDERLYING THE NEW LEARNING AND TEACHING STRATEGIES

There is no direct simple correlation between the content of engineering education and the functions and responsibilities of an engineer's professional activity. Nonetheless, the modern engineer should be :
- a professional with a broad scientific background capable of mastering the complexity of new technological fields, and/or of industrial, financial and economic organizations;

- an entrepreneur, creative and aware of his/her responsibility to society so as to be a locomotive in helping it evolve;
- an accomplished communicator with the capacity to listen, to discuss differences, to convince and to inspire the team he/she works with:
- open to international realities, ready to face competition (the essential motor of growth) every day in technological, industrial and commercial fields.
- desirous of constant improvement through continuing education;
finally, both modest and moral.

## 4. THE NEW CURRICULUM

The aim of Ecole Centrale Paris is to produce top-level, multi-disciplinary engineers and managers. While in constant evolution to anticipate and reflect the needs of the fast-changing industrial and economic environment, the three-year curriculum at Ecole Centrale Paris consistently emphasizes a balance between three fundamental aspects:
- acquisition of a broad, multi-disciplinary scientific and engineering background
- development of interpersonal skills
- introduction to the dynamics of the engineering profession.

Note again that all students come to Ecole Centrale Paris after at least two years of post-high school education. During their first two years at ECP, all students follow a common core curriculum. Then, in their third year, they choose an area of concentration among fifteen engineering fields.

*4.1. The « Common Core »*

*4.1.1. A broad scientific and engineering background*

Two-thirds of the students' time is devoted to the acquisition or reinforcement of scientific and engineering knowledge. The program combines a variety of formats varying from lectures and tutorials to laboratory and project work. A prime characteristic of the academic program at ECP is its multi-disciplinary focus. Students gain exposure to a broad spectrum of disciplines ranging from most engineering fields to economics and management. They also acquire the scientific tools that are useful in these fields. Priority is given to acquiring fundamental knowledge and methods, developing an understanding of the issues and concepts of importance to the various fields, rather than delving into technological applications that may rapidly become obsolete. This approach is based on the double premise that the students admitted to Ecole Centrale Paris have the intellectual abilities to rapidly adjust to the technological tools specific to their future career field, and that industry understands that it will provide the necessary complementary technological training to make their future employees operational.

This original approach is validated by the tremendous success encountered by ECP alumni in the very diverse fields they enter, whether in engineering, research, or management. It should be noted that Ecole Centrale Paris has been always considered among the top three schools in France with respect to employer satisfaction. In 1998 and 2000, the "Nouvel Economiste" ranked Ecole Centrale Paris at the top of the list for employer satisfaction.

*4.1.2. Introduction to the Engineering Profession*

*The Integration Cycle :* This cycle helps the students to become aware of the business world and the school's educational objectives so that they will modify the frame of reference attached to their previous studies. The first semester offers several opportunities upon which they can discover how companies function: the half-day presentation of companies, the Integration Weekend, the half-day presentation of research activities, the visits to factories, the meetings or round-table discussions with young practicing alumni, etc...

*Introduction to the corporate world and interaction with industry :* Whatever their future profession, high-level engineers and managers strongly benefit from an in-depth understanding of the industrial world. The faculty is a major element in this area, as more than half of the teaching is done by outside part time professors, most of them with their main activity in industry or services. Several activities (spotlight study, team project...) are done in conjunction with industry, and a non-specialized internship is compulsory.

*Learning doubt: developing problem solving capacities :* In order to transmit knowledge as quickly as possible, teaching is generally done deductively, which reinforces the students' certainties. Reality is quite different. To expose students to the beneficial effects of questioning, they have project work which gives them an initiation to research (research projects, industrial projects). To consolidate their knowledge, the dimension of group work has been incorporated into the curriculum, and team projects are offered to each class. These correspond to the characteristics of a real-life situation with a need expressed by a project manager, a defined objective, and resources allocated to reach the desired ends.

Students thus discover the importance of bibliography, the difficulty of experimental work, the interest in protecting innovation, the efficiency of teamwork, and the other aspects linked to fulfilling a contract through to the end. This provides them with an opportunity to participate in the activities of the school's Research Laboratories. All graduating students will thus have had contacts with research, in addition to the laboratory sessions which also take place there.

*4.1.3. Development of personal and interpersonal skills*

*Social and Human Training :* Most students entering Ecole Centrale Paris have had little prior opportunity to familiarize themselves with the diverse aspects of engineering. Several social and human sciences are offered at ECP (including a course in the philosophy of sciences and a business simulation game) to give students a better understanding of themselves, as well as of interpersonal relations and group dynamics.

*Awareness of Other Cultures* : The presence of foreign students on campus (20% of the student body), the possibility of finishing their studies abroad and obtaining a double degree (more than 25% of the graduating class), the teaching of language by native speakers, the obligation of reaching a high level of competence in the two languages studied, and the mandatory internship abroad in a non-French speaking country all contribute to the students' acquiring this awareness.

*International programs* : Ecole Centrale Paris is a founding member of the TIME network, which involves double degree exchanges among more than thirty European universities. Each year, Ecole Centrale Paris receives some 60 European students for the Common Core Curriculum. They study on campus for two years of multidisciplinary engineering education, then return to their home institution to complete their education. The other students may study in the TIME network in place of their final year, in which case their studies are generally lengthened by one year. They may also study abroad in the United States, Canada, Great Britain or Japan. Through agreements, they are accepted for a Master's degree in place of their third year at ECP. In 2000, 85 French students left Ecole Centrale Paris for foreign Universities.

### 4.1.4. A Personalized Education

*Elective modules* : Elective « thematic modules » have been proposed in the six thematic groups :
- Mathematics
- Physics
- Mechanics
- Engineering Sciences
- Electronics, Control Systems, and Computer Science
- Economics and Business Management

The modules give the students an opportunity to acquire more specialized knowledge on a chosen subject in each field. Attendance is mandatory, and the pedagogical activities proposed by the professors vary greatly.

*The Log Book* : All students are given a personalized Log Book which sets forth the seven objectives they are expected to attain through their various activities:
- acquire broad competence in scientific and technical fields
- grasp complexity
- understand the basics of company organization
- develop a spirit of initiative, a critical sense, and an aptitude for innovation
- increase personal development and cultural awareness
- develop communication skills and teamwork
- recognize and understand the importance of other nations and cultures

Reaching each of these objectives is divided into stages (scientific, technical and methodological competence and/or relational and interpersonal) that are validated over the two years of the Common Core, each noted in the student's Log Book. A certain number of optional activities make it possible for the students to personalize their study plan while meeting the requirements for the validation of their curriculum. Extracurricular activities can be taken into account.

Ecole Centrale Paris cannot of course claim, or even hope, to transform each of its students into an « ideal » engineer who has reached the mastery of all of the objectives enumerated.

Nonetheless, Centrale's duty is to place its students in situations which allow them to confront these aims through examples illustrating their importance. The intention is to set in motion a process of learning that the future engineer will use throughout professional life in numerous concrete situations.

*The Tutors :* Throughout the common core curriculum, each class is divided into groups of twelve students, with two tutors drawn from among the professors, administrative staff and Ph.D. students. Each tutor has three private meetings a year with six of the students in the group, the aim being to help the students make their choices and overcome the difficulties they face. Students undertake several activities in the context of their tuition group, including the « Spotlight Project » which focuses on a technically complex object.

*School Regulations : Initiative, Autonomy and Responsibility :* A list of the school regulations is given each year to all incoming students. Students must meet the academic conditions in the Student Manual and validate the stages defined in their Log Books. This said, they are allowed a maximum amount of freedom in the organization of their time.

*4.2. The third year areas of concentration*

The third year at Ecole Centrale Paris is designed to provide depth in a particular area of concentration of the student's choice. The area of concentration is not intended as a specialization (it does not appear on the ECP diploma) and the students do not necessarily begin their professional careers in the corresponding field. The differences between a specialization and an area of concentration may perhaps appear subtle. The objective of the area of concentration is to develop adaptability and flexibility. The training provided in an area of concentration simulates the industrial environment wherein the newly-hired engineer must acquire proficiency in the activities of the company over a very short period of time.

Each area of concentration is in strong interaction with industry : Nearly one hundred French and foreign companies sponsor the final year and participate both in the teaching and in the orientations.

There are currently about 15 areas of concentration offered in a wide range of engineering fields that reflect the multi-disciplinary character of the school. The organization and balance between lectures, tutorials, laboratory classes, projects, and factory visits may vary significantly between the various areas of concentration. Nevertheless, the common guideline is to develop methods for efficiently and rapidly acquiring in-depth knowledge and experience in a given field.

The third year of studies at ECP is divided into two periods:
- **Formal training** (6 months, September-February): Lectures, tutorials, laboratory classes.
- **Internship** (3 months, March-June, or longer): Either in industry or in a research laboratory. The internship culminates in a written report with oral defense.

A new (experimental) area of preparation to the engineering profession, "Management of Industrial Projects", was started in 1997. This area is a personalized cross-disciplinary program with a total duration of 16 months that a student can choose in parallel to his or her area of concentration.

A student may also, if he/she intends to begin professional life by creating a start-up, follow a personalized preparation including the help of business angels. Thus, six start-ups began their life in 1999-2000.

## 5. THE ONCOMING CHANGES

They derive essentially from an extension of the work done to improve the Common Core, so they concern the third year curriculum.

Trying to imagine what a young engineer graduating after the area of concentration should be has led to the conclusion that he/she should at the same time:
- Deepen his/her knowledge in a given disciplinary field
- Apply this knowledge to a given industrial sector
- Be conscious of the bases of one aspect of the engineering profession

To achieve this, and starting in September 2001, the third year curriculum will be modified and the 1997 experiment extended to all students:
- Eight areas of concentration focused on a set of **Disciplines** essentially oriented towards an industrial **Sector**: Mechanical Engineering, Civil Engineering, Process and Environmental Engineering, Computer Science, Electrical Engineering, Applied Physics, Industrial Engineering, Applied Mathematics.
- Five areas of preparation for one aspect of the engineering **Profession**: Project Management, Finance and Consulting, Design and Research and Development, Production and Logistics, Business Start-up.

Each student will choose one area of concentration (seven months of formal training) and one area of preparation for one aspect of the engineering profession (two months interwoven with a seven-month long internship related to both areas).

## 6. CONCLUSION

Since its foundation in 1829, Ecole Centrale Paris has not changed its major objective: training high level multidisciplinary engineers for industry. Yet, it has constantly changed its curriculum in order to accompany the evolution of the world.

Today, an engineer cannot not only be a specialist in Science and Technology, but he or she must complement his or her scientific and technical aptitudes by attitudes that derive from an understanding of human and social sciences, and should in addition be well aware of industrial and international realities. The new curriculum at Ecole Centrale Paris has been built in order to integrate all these factors and progressively lead the students towards initiative, autonomy and responsibility.

*E. Esposito, E. Sigler*
*Ecole Centrale Paris*
*Paris, France*

V. SINHA

# INTERDISCIPLINARY CURRICULUM FOR DEVELOPMENT OF A GLOBAL ENGINEER

**Abstract.** Requirement for mobility of engineers across the national boundaries has changed the traditional demand-supply perspective necessitating recasting of the engineering curricula. The engineers of today must adapt themselves to ethos of the various societies where they would work, must be concerned with the local environmental issues, etc., besides being good in their fields of work. It is essential that the training that is imparted to them should prepare them to these added challenges. We present a blue print for such a curriculum. The focus is on undergraduate studies.

## 1. INTRODUCTION

In the present global scenario with free flow of technology, trade and commerce we require adequately trained engineers to serve the economies across the nations. Today we have business and management schools (graduate schools) whose products are shaping the destinies of the corporate world. But, it is not enough to have engineers who have specialized in some management streams to take care of the truly global scenario. We need to evolve engineering curricula, using interdisciplinary approach, which will also lay stress on the cultural bias of the society as well as the local environmental issues besides the economic as well as industrial practices. Integration of these subjects in the traditional curricula could only be at the expense of some of the existing topics in the present set-up, given the constraint of four/five years (eight/ten semesters) of undergraduate engineering education. We propose a blueprint for such a curriculum based on the following ingredients:

(1) Input quality assurance

(2) Interdisciplinary curriculum; with stress on economics, infrastructure management, societal (humanities and social sciences) and environmental concerns

(3) Excellent faculty and infrastructure

(4) Exposure to industrial practices

(5) Periodic review and upgradation based on identification of national needs/international trends,

(6) Assimilation of the impact of Information Technology

Part of the pedagogy could be through remote tutoring. It is expected that summer internships would be a crucial aspect of the curriculum. Such a curriculum requires very close interaction of the academia with the industries. Any curriculum design is a dynamic process and the contents of the curriculum would change with the passage of time. Also, there are large numbers of changing and fuzzy parameters that influence the curriculum. In the present scenario, with rapidly changing technologies, the design must incorporate the impact of such changes.

## 2. CURRICULUM DESIGN

No curriculum can ever be designed which will fulfil the requirements of different kind of industries in all respects. A fresh graduate will not be productive from the day he/she joins an industry. Certain amount of 'on the job' initiation is essential. What is required of the curriculum is, whether the student has been given adequate background knowledge to understand the concepts of the discipline as well as the needs of an industry. The curriculum should inculcate in the individual the ability to think, analyze and decide on the tasks before him. Horizons of technology have always been expanding; the present decade has seen the expansion process growing exponentially. The curriculum should impart the training to assimilate the changes in technology. Having said that, as an existential theorem, it is really difficult to develop a global blue print to achieve the objectives.

A curriculum affects three constituents: students (who are the prime movers), faculty (who are responsible both for designing and for overseeing that the curriculum is administered) and the industry/society (who are the beneficiary from such trained students). There should be synergy amongst the three constituents as to what should go in the curriculum. There would be issues wherein there will be resonance amongst the three and there will be issues where there would be conflicts amongst the three. A student is interested in earning a respect (from the knowledge that he acquires) as well as a job for the furtherance of the goals of his life. Faculty is interested in disseminating knowledge (and only through such process can she/he generate more knowledge). The industry is interested in immediate utilization of the knowledge that a student has acquired for furtherance of its objective of creating wealth. Naturally, there could be areas of commonality and areas of conflict. The curriculum has to have a balance for the benefit of all the three constituents. The objective of the curriculum should be to produce engineers who would have
- versatility – ability to meet the challenges due to changing technologies,
- competency in the discipline,
- life-long learning enthusiasm,
- societal concerns.

Even though one may like to re-think about the curriculum from a scratch, one cannot disown the past practices and the principles. After all, curriculum gets evolved over a period of time, based on inputs from all quarters and from the passage of time. Pedagogy is about dissemination of knowledge. One builds curriculum on the platform, in our context; it is on the level at which students as inputs to the pedagogic process arrive. Curriculum may differ according to the levels of inputs. Therefore, each system will have to update (by continually changing the emphasis – based on the perceptions of the faculty at the time of change – and by introducing new material as the technology brings). For example, one may like to discard the rigour of analysis to introduce newer concepts. While designing curriculum, often-asked questions are answered below.

*2.1. How much "basic knowledge" is enough ?*

No amount of "basic knowledge" is enough. It is a compromise that the designers of curriculum have to incur. The basic knowledge in the context of, say, Electrical Engineering (EE), would not only include Physics, Mathematics, Computation, Technical arts, Engineering sciences but also will include environment, materials, humanities and social sciences, besides, of course, the subjects of Electrical Engineering. The issue to be addressed to would be: given the constraints of an 8-semester 15/16-week course-work, what quantum of time will be apportioned to different subjects. It is a jigsaw puzzle with many pieces fitting to make the picture. There cannot be a unique solution.

*2.2. What balance to strike between "the broad picture" and specialization?*

It is imperative that an EE working in say electromagnetics has sufficient knowledge of electrical machines; at least at the base level. Hence, he should have strong fundamental knowledge in all the disciplines of electrical engineering: circuits, electronics, signal processing, communications, controls, power systems, power electronics, electromagnetics, devices etc. At the same time, he should have adequate knowledge in his specialization.

*2.3. What balance between practice and theory?*

Ever since the computers have invaded our class-rooms (should we say our lives !) the emphasis on hands-on experience or laboratory skills have dwindled. This is not a healthy trend. After all, a society will need all kinds of engineers to design, fabricate, maintain and innovate whole gamut of, say, electrical/electronic/ communication systems, machinery, processes both in hardware and in software (assuming, again, that the curriculum if for electrical engineering). There cannot be a short-cut to hands-on experience and therefore, laboratory skills must be emphasized.

Based on the discussions above, we come to realize that the courses in the curriculum have to be designed to fulfil the following three fundamental tenets:

- sharpening of the mind (through courses in sciences, especially in mathematics and logic),
- developing the concepts – strong fundamentals in the broad discipline with in depth knowledge in specialization,
- developing the skills – through training imparted through state-of-art laboratories.

## 3. A SUGGESTED MODEL

A suggested model for development of curriculum is given in Table 2. The table indicates possible break-up of courses. We have, once again, assumed that the curriculum is for Electrical Engineering discipline. The sub-heading of Humanities/Social Science courses include the courses on environmental awareness and management strategies.

*Table 1. Break up of Courses*

| Major Headings | Percentages |
|---|---|
| Science courses (Mathematics/Physics/Chemistry) | 15% |
| Humanities/Social Science courses | 15% |
| Engineering Sciences courses | 10% |
| Technical Arts | 5% |
| Non-EE electives | 12.5% |
| EE Compulsory Courses | 17.5% |
| EE Electives | 20% |
| Project work | 5% |

Such a distribution ensures freedom to a student in widening her/his horizon by taking elective courses from outside the discipline of EE and also in specializing in a field of her/his interest where she/he might like to spend working rest of her/his life. The training imparted through the curriculum should be augmented by industrial training, during the semester breaks, preferably in other countries. It must be made mandatory to expose the students to the industrial practices as well as for cross-cultural experiences.

Any curriculum assumes certain basic level of input. The output depends on the quality of input as well as on the level of faculty and the infrastructure for the engineering programme, in terms of state-of-art laboratories, library resources, networking and computing facilities. Whereas input quality could be ensured by a screening test for students to be admitted to the engineering courses, based on sound preparatory knowledge in science subjects and aptitude for engineering, quality faculty and good infrastructure are difficult to provide. Most of the developing countries are unable to attract good faculty and provide adequate resources for infrastructure. Augmenting the gap for regular faculty by remote tutoring could be a solution but the network bandwidth issue, still, remains. Similarly industry-academia interaction are to be strengthened.

## 4. CONCLUSION

An attempt has been made to present issues relating to the engineering curriculum design which incorporates deserving emphasis on courses relating to societal concerns and environmental issues. It is argued that the curriculum must emphasise on the training which sharpens the mind, develops sound concepts of the discipline as well as provides sound engineering skills.

*Vishwanath Sinha*
*Indian Institute of Technology Kanpur*
*Department of Electrical Engineering*
*Kanpur, India*

R. L. KING, S. MELSHEIMER, R. MOSES

# A MULTI-UNIVERSITY ENGINEERING SUMMER STUDY ABROAD PROGRAM

**Abstract.** This paper will describe a successful engineering study abroad program that occurs every summer for six weeks. The program participants are engineering and science students from Mississippi State University and Clemson University in the United States and the program takes place at the University of Bristol in the United Kingdom. The program is designed to handle approximately 30 students in total from the two US schools and permits wavering enrollments at one university to be compensated by the other. Thus, the University of Bristol can be assured of a reasonable number of students on a continuing basis to keep the program profitable. The first four weeks are spent at the University of Bristol, where students take two classes and earn six credit hours. The program ends with a two-week independent travel period where students typically tour the continent. While attending classes at the University of Bristol, the students experience British culture firsthand by living with host families.

## 1. INCEPTION OF PROGRAM

The first attempt at beginning an engineering summer study abroad program began in 1977 when Wayne Bennett tried to interest Virginia Tech students in studying at the University of Leicester during the summer of 1978. At this time a rule of thumb was developed that stated that it would take fifteen students to make the program successful. Finally, in the summer of 1980, nineteen students travelled to the University of Leicester. However, the following year Wayne Bennett moved to Clemson University and the Virginia Tech program converted into a one week Spring Break program.

## 2. UNIVERSTY OF BRISTOL PROGRAM

### 2.1. Program Description

The summer study abroad program is conducted during the month of July at the University of Bristol in the United Kingdom. This corresponds to the second summer term at each of the US schools. This is important because it permits students who are on cooperative education and work programs to attend their US school the first summer term. Courses are taught by faculty from the University of Bristol and all lectures are in English. These two parameters (timing and language) are essential in maintaining a consistent enrollment.

Bristol was selected not only for its well-known engineering school, but also because of the science and technological history of the area. The City of Bristol is almost 1000 years old and is one of the most attractive large cities in Europe. The Bristol area has a tradition of engineering achievements and remains on the frontier of engineering development with many high-tech companies including British Aerospace, the builders of the Concorde, and Rolls Royce aero engines. Bristol is also the center of Britain's version of Silicon Valley with such companies as Hewlett Packard and Du Pont having factories and research centers near by.

The program does not involve transfer credit. Both US universities have their students enroll in course numbers at their home university with course content equivalent to that to be taught at Bristol. This facilitates the use of student financial aid (scholarships and loans) since students cannot apply for financial aid for transfer credit. Students will enroll in two courses worth six credit hours, which is a full load for a summer term at both of the US universities. Also students participating in the program will normally have a Junior class rank. This ensures they have the mathematics and sciences background to be successful in the program.

The first course is a humanities course reviewing the history of science and technology. All students attending the program are required to take this class. The course's focus is primarily the industrial history of the Bristol area. The course pedagogical approach is a mix of lectures from noted scientists and engineers from the area and field trips to selected museums and industries. Assessment of this course is primarily based on papers written by the students on a variety of topics that explain how technology has impacted the growth of an industry. The University of Bristol faculty member in charge of the course assigns the final grade.

The second course serves as a senior technical elective for the study abroad students. Most US universities permit their students to select from a variety of technical courses to complete their degree requirements. This flexibility allows students to specialize in a technical area in which they might desire to seek employment upon graduation. Originally, there was only one technical elective course offered in the program, but recently due to higher enrollments in the program a second course has been added.

The principal technical elective course has two thrusts associated with it – 1) mathematics of computer graphics and 2) robot manufacturing. This course has been offered since the program inception as a course that introduces science and engineering students to the mathematics of computer graphics and robot manufacturing. The computer graphics pedagogical approach is a series of lectures by a University of Bristol faculty member of the mathematics (spline curves, etc.) behind CAD/CAM programs. The assessment tool is a three-hour written examination at the end of the study to determine the student's comprehension of the material.

The second aspect of this course is a laboratory-based introduction to robot manufacturing. The students are divided into teams of three and are required to take a concept for a chess piece from an initial design through to manufacturing. The students meet as a team to decide on a design for a chess piece, develop a 3-dimensional CAD/CAM design, determine a manufacturing process, program the robotic machines, simulate the process, manufacture the piece, and prepare a written report on their project. Two machines, a robotic mill and a lathe, are available for the student's use in manufacturing (see figure 1). An assessment of the team's work is then determined based on the written report. A final grade for this course is then assigned by combining the student's marks from the mathematics examination with the written laboratory manufacturing report.

*Figure 1. Students manufacturing chess piece using robotic machines.*

In recent years, due to increased enrollment from Mississippi State University students, a second technical elective course has been added. This course is a digital signal processing course. It involves both a lecture and laboratory component. This course is limited to students studying Electrical Engineering, Computer Engineering, or Computer Science. This limitation is to ensure students have the programming skills necessary for the course material. The course utilizes hardware boards for signal processing and requires students to learn how to program and use the boards in practical applications. The assessment is determined based on student marks from both written examinations and laboratory work.

*2.2. Clemson University*

The summer study abroad program started in its present form between Clemson University and the University of Bristol during the summer of 1985 when twenty-five students travelled to Bristol. The program continued with only Clemson participation through 1996. Table 1 shows the student enrollment during that time period. Note that the program sent students abroad for 8 of the 12 possible summers. Also, note the difficulty in maintaining the minimum number of 15 students.

*Table 1. Clemson University participation during the 1985-1996 period*

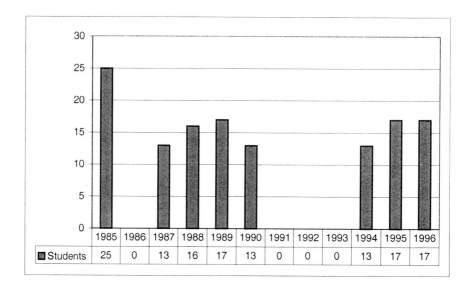

### 2.3. Mississippi State University

During the summer of 1997, Mississippi State University joined with Clemson University in the summer study abroad program. The benefits of this alliance were twofold. First, it permitted Mississippi State University to immediately have access to an on-going program with a European school and with courses already in place. This permitted Mississippi State University to model its program on Clemson's and be able to immediately send students abroad. For Clemson University, it brought in another source of students to ensure the continuity of this program.

Table 2 shows the combined university enrollment figures for 1997-2001. It was when enrollment reached the 30 student mark (1999) that the University of Bristol added the second technical elective course – Digital Signal Processing. This was required due to the fact that the robotic manufacturing laboratory was limited to a maximum of about 21 students.

Also, this alliance has ensured the viability of the program for the future because the University of Bristol has recently raised the minimum number of students required for attendance to 20. The average student participation prior to 1997 was 16.4 students per summer. After the alliance between the two US universities it has now jumped to 24.6. It is also important to note that during this last five-year period that the program would not have been viable without the alliance between the two universities.

*Table 2. Clemson and Mississippi State University participation during the 1997-2001 period*

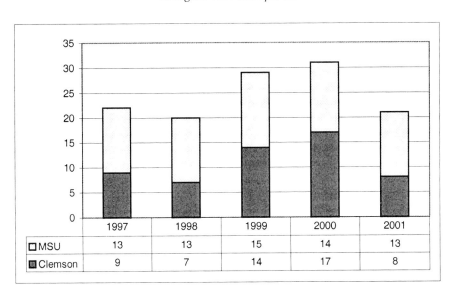

Finally, Table 3 combines the student participation statistics form the inception of the program at the University of Bristol.

*Table 3. Student participation during the 1985-2001 period*

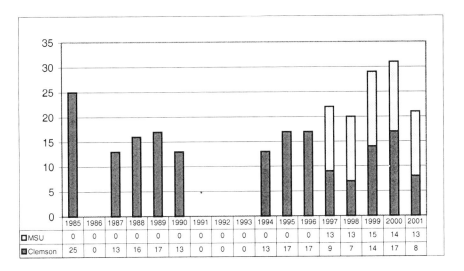

## 3. CULTURAL EDUCATION

An important aspect of the program is the non-technical (i.e., cultural) education it provides to the students. What has been one of the most successful aspects of the program is the home-stay while the students are studying at the University of Bristol. Students are matched with families within walking distance of the university so American students can truly experience the local culture of the community. In addition to a place to sleep and study, students have two meals per day provided by their new family. Students are encouraged to develop interactions with their host families and not just be a boarder. In post-travel assessments this aspect of the program has always been singled out as one of the most meaningful parts of their study abroad experience.

All the students participating in the program will travel one day to Stratford upon Avon to view a Shakespearean play and have time to explore the town of his birth. Students also have the opportunity to travel on the weekends beyond the local area. This occurs in one of two ways – British Rail passes or on coach excursions hosted by the University of Bristol faculty. The Mississippi State University students are provide with four day British Rail passes as a part of their tuition funds and are encouraged to travel to areas around the United Kingdom that they might otherwise miss (e.g., London, Edinburgh, York). Also, the university faculty hire coaches for one-day excursions during the weekends to such local areas of interest as Stonehenge or the castles of Wales.

Finally, after the four weeks of formal education at the University of Bristol the students have two weeks of independent travel around Europe before returning to the US. For safety reasons, students are encouraged to travel in small groups and to agree upon basic itineraries that they give to the University of Bristol faculty before departing. They are also provided with emergency contact numbers for people in the United Kingdom.

Students are encouraged to procure their independent travel rail passes before departing to the United Kingdom for their studies. Also, to facilitate their independent travel, the students may leave many of their clothes and other items in a secure location at the University of Bristol. This provides the students the ability to travel with backpacks or other lightweight luggage during their independent travel. Upon returning to London to return to the US, the university will return their luggage to their local hotel in London before departure.

## 4. CONCLUSIONS AND RECOMMENDATIONS

The University of Bristol has been the host university for a summer study abroad program since 1985. The program began with only Clemson University as the US participant, but due to maintaining enough students in the program Mississippi State University joined the program five years ago. The lessons learned from this experience are many. To ensure program viability the following recommendations are given:

1. Have multiple US universities participate in the program with a European host university.
2. US universities must have compatible semester time schedules.
3. English language is important for course work, but not essential for non-university aspects of the program.
4. A home stay versus a dormitory is preferred.
5. Courses offered by the host university need to be a combination of technical and non-technical alternatives.
6. Provide for some cultural activities during the study portion of the program.
7. Allow time in the program for students to have independent travel.

*Roger L. King - Mississippi State University, U.S.A.*
*Steve Melsheimer – Clemson University, U.S.A.*
*Roger Moses – University of Bristol, U.K.*

M. RÖTTING

# INTEGRAL - A WEB-BASED TOOL TO SUPPORT LEARNING IN INTERDISCIPLINARY TEAMS

**Abstract.** This paper describes the INTEGRAL system, a web-based learning tool to support university teaching in the area of ergonomics / human factors. INTEGRAL is an interactive, multimedia, and location-independent learning environment supporting traditional lectures and exercises. The interdisciplinary and cross university co-operation of students, e.g. in project groups working together on a case study, is an important aspect of the INTEGRAL system. The INTEGRAL system offers various modules (e.g. lecture materials, multimedia files, case studies, study projects, communication forums). Teachers can define a filter function selecting those modules that are appropriate for their students' needs and their didactical concept. A preliminary investigation pointed out the requirement for such a training system. The acceptance of the system was evaluated with a prototype of the system focusing on cross university collaboration between students from Aachen and Essen. It was shown that the INTEGRAL system can well motivate students. However, it was shown too that new means of communication alone are not sufficient and that a combined approach with classical means of communication (e.g. accompanying seminars) should be preferred.

## 1. INTRODUCTION

Teaching ergonomics / human factors at German universities is characterized by students coming from various faculties. E.g. students from different fields of engineering, from product design, business sciences, psychology, pedagogy and others are attending the lectures. For some students, ergonomics is a required course, for others it is one out of a number of selectable courses. This underlines the importance of the subject, but results in difficulties in presenting the subject adequately in more traditional, frontal forms of presentation. In addition, what is standard in today's industry, the collaboration of professionals from different domains, is seldom found in university teaching. Student collaboration in interdisciplinary teams and teams from different universities is the exception.

Especially demanding to teach are the design relevant parts of the ergonomic knowledge. The teaching subject is multi-disciplinary. Knowledge from different disciplines is needed to solve ergonomic design problems in industry. It is difficult to transfer available knowledge and previous experience to the problem at hand to solve it.

The main focus of the project INTEGRAL was therefore
- the development of an interactive multimedia course content to strengthen the students ability to solve design relevant problems in product and workplace design.
- to enhance the communication between students of different fields of study,
- to practice goal oriented use of new media, and
- the development of an efficient platform that can be used for similar teaching situations. Therefore the program code and the domain specific content are kept separate.

The project groups consist of four university partners from the University of Technology Aachen, Essen University, Siegen University and the Institute for Occupational Physiology at the University of Dortmund. All four are from the German state North Rhine-Westphalia, whose government funded the project.

## 2. PRELIMINARY INVESTIGATION

*First questionnaire*

One of the first steps within the project was the determination of the demands of the students regarding such a system. A questionnaire was developed compromising questions regarding
- the use of a PC and the Internet,
- the importance of new media in general and applied to university curricular,
- the accessibility of PCs and the Internet now and in the future,
- the courses in ergonomics,
- the use of the Internet as part of the ergonomic courses and as means of communication with other students,
- the estimation of effort involved in web based training, and
- personal data: age, gender, field of study, number of semesters and the relative importance of the courses in ergonomics.

The questionnaire was administered to students in Aachen and Essen during the 1998/1999 semester. A total of 61 questionnaires was returned and analyzed.

*Results of the preliminary investigation*

The results of this first questionnaire in general supported the concept of a web based learning tool. The students have a positive attitude towards new media. Almost all of the students (97%) say it is important to become familiar with new media and 72% of the students request a wider use of new media in university education. Only 18% of the students have reservations regarding the use of new media in school and further education. And 62% regard the Internet as one of the most important information source of the future.

It was important to find out if the students have experience in using and have access to a PC and the Internet. Here again the results were supportive: 92% of the students use a PC at home and 56% more than five hours a week. Even at the time of the administration of the questionnaire, 72% of the students had access to the Internet and the majority of those even from their home (61% of all respondents).

Regarding the use of the Internet in Ergonomics / Human factors lectures, the following items were rated as important or very important by the students:
- Independent compilation of ergonomics / human factors specific content by 55% of the students.
- Individual consolidation of personal interests within the field of ergonomics / human factors (87%).
- Free timing (67%).

- Working on case studies (77%).

The questionnaire mentioned a number of different potential fields for the application of new media in university education. The students agreed to them to different degrees: 79% of the respondents welcomed the possibility to access supplementary material via the Internet, 67% of the students welcomed the possibility to get lectures on demand via Internet and 53% welcomed the possibility to work on exercises and theses via Internet. Only 15% welcomed the possibility to have examinations via Internet.

If lecture material is presented on the Internet, 50% of the students expect that they would spend more time than attending a traditional lecture. In addition, 48% of the students believe that more self initiative would be required. To work together with other students from different faculties and universities can be envisioned by 82 % of the respondents.

Summing up the results, the demands and wishes of the students do well agree with the goals of the project. The points seen critically by the students, e.g. the possibility to have exams via internet, are not intended by the project. On the other hand, the high acceptance of working together in cross-disciplinary and cross-university teams is supporting the concept of INTEGRAL.

The students do have a realistic expectation regarding higher demands on their time and self initiative. A study by Scheuermann (1998) showed that basic skills to use the Internet are needed, that more self discipline is required and that working in a group needs to be trained. But it is expected that these higher demands are balanced by the acquisition of additional competencies, e.g. media and communication competence.

*Didactic concept*

Starting point of the learning with new media must be a didactic concepts that allows to combine the advantages of the different media. Figure 1 shows the didactic concept for the INTEGRAL system. It is based on the recommendations regarding the design of multimedia teaching systems by Beitinger & Mandl (1992), Harms (1998) and Unz (1998).

Multimedia teaching systems should be part of an integrated learning environment to utilize the specific advantages of the different media. They demand supporting activities during all phases of the didactic concept. New information and communication technologies do not make classical educational concepts obsolete, but amend them and enhance their efficiency (cf. Schwuchow, 1997). Additional activities like workshops and seminars can combine the advantages of face to face teaching (e.g. higher motivation, better dealing with problems, social interaction) with the advantages of remote instruction (e.g. independence in time and space) (cf. Geyer et al., 1998).

Such a didactic concept can be visualized in four phases (cf. Schwuchow, 1997; Stockfisch & Sigel, 1997):

- **Preparation:** This phase compromises the preparatory measures like the development of different seminars for teachers and introductory seminars for the students, the design of the software for the web based learning tool and the development of Internet pages for the case studies.

- **Knowledge mediation:** The students have to acquire basic theoretical knowledge. This knowledge is brought to the students e.g. in lectures. It needs to be reflected and evaluated regarding it's relevance to solve the case study the students work on in teams. Additional information is needed to solve the case studies and the students have to find this information by themselves, e.g. by an Internet search. The theoretical knowledge is therefore applied in a practical context. The case studies require the cooperation in interdisciplinary teams.

- **Consolidation:** In this phase the experiences gained will be exchanged in a workshop. Especially the processes of learning and working together in a team will be discussed. In addition, the intermediate results of the case studies will be presented. So, the results and the way of reaching them have to be reflected by the students.

- **Application:** The goal of the INTEGRAL project is the strengthening of media competence and of methodological competence to better solve applied problems in the field of ergonomics. Computer networks are not only used to train the students. They are at the same time tool for solving a problem and object of analysis. This creates an authentic context for knowledge acquisition. The students have to solve a problem by themselves and their communication competence is strengthened (cf. e.g. Fischer & Hägebarth, 1999). The discussion with other members of the group requires a reflection of ones own position and it's theoretical basis. This is how multiple views of the problem are generated. The asynchronous technology allows an intensive elaboration without time restrictions (cf. Harms, 1998).

The four phases preparation, knowledge mediation, consolidation and application should not be seen as a simple chronological sequence. Indeed the four phases follow after each other in a cycle that has been named Learning Cycle (Mayes et al., cited by Geyer et al., 1998).

# INTEGRAL - A WEB-BASED TOOL TO SUPPORT LEARNING

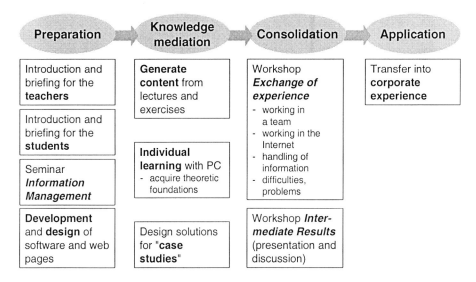

*Figure 1. Illustration of the didactic concept of the INTEGRAL project.*

*Concept for the learning tool*

The INTEGRAL system is offering material in addition to the traditional ways of teaching (like lectures) and offers tools for the collaboration of groups of students. All material is arranged in modules. New modules can be added over time. Since the INTEGRAL system is part of the Internet, a collaboration of students from different universities is easily achievable.

Each teacher will still follow his or her own didactic concept. INTEGRAL can easily be adapted to this differing demands due to the modular structure. As can be seen in figure 2, the pool of offerings – the different modules – are made available to different users via a filter. The filter can be defined by each teacher for different groups of students according to the didactic demands. Students can be members of more than one group and additional modules can be made available on an individual basis as well. To work with the INTEGRAL system, the students have to logon to the system with a user name and password.

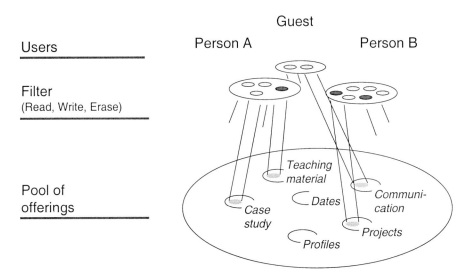

*Figure 2. User specific selection of offerings.*

## 3. PROTOTYPE, QUESTIONNAIRE AND CONSEQUENCES

*Prototype*

A first realization of the INTEGRAL system was build. It is web-based and therefore accessible by all computers with a graphical browser. The content is placed in a SQL database with a tree-like structure. The filter allows to restrict the access to different branches of the tree. Students can have read only or reading and writing rights for different areas of the system. In addition, they might have the right to build new branches. Every piece of information can be conceived as a message. Depending on the rights, every user can comment messages, add messages or even start new branches of messages. Within each message there can be text, graphics, multimedia content and links to other parts of the INTEGRAL system and the Internet.

This first prototype of the INTEGRAL system was evaluated by a group of students from Aachen and Essen. They had to collaboratively work on a small case study (in addition to the conventional lecture / as an additional workshop). The goal of this evaluation was to
- test the acceptance of the systems as a future means of teaching,
- get input for different design variants of the interface,
- analyze the user interaction, and
- get hints for the future development of the system.

## Evaluation of the interface

The evaluation of the interface showed that at first not all students could access the system in a satisfactorily way. It is important that such a system must function with many different hardware and software configurations. But otherwise the evaluation proved that the navigation is comprehensible and coherent. The „bookmark"-function was very well received. In addition, the students gave many hints regarding the elements of the screen design (e.g. notations, symbols).

## Results of the questionnaire

The participating students had to answer a questionnaire. It compromised questions regarding motivation, team work, the teachers, use of media, time needed and other aspects of system use.

The results regarding motivation are very promising. The internet based work was conceived as interesting, not tedious and fun. The work in a team is more fun, multifaceted and more interesting than more traditional forms of teaching.

The aspect of working together in a team needed some further consideration. The definition and allocation of tasks was perceived as bad. The timely settlement of tasks, the making of arrangements and fulfillment of dates and arrangements was rated as only fair. But the distribution and provision of information was rated as good.

The use of media is shown in the following table 1.

*Table 1. Use of media and information quality*

| Rank | within Team | with teacher | Information quality |
|---|---|---|---|
| 1 | email (29%) | email (32%) | personal (51%) |
| 2 | chat (21%) | personal (team) (32%) | email (21%) |
| 3 | personal (team) (13%) | phone (8%) | photocopies (10%) |
| 4 | phone (12%) | chat (6%) | phone (5%) |
| 5 | personal (2) (9%) | personal (2) (5%) | news (4%) |
| 6 | news (4%) | | chat (4%) |

Within the team, 54% of the communication was done using new media. With the teachers, this rate was only 38%. The rating of the quality of the information gives a clear preference for personal face to face communication. The students do not prefer internet based media over classical media.

## What does that tell us?

The INTEGRAL system can very well motivate students. The aspect of working together in a team needs further consideration. The team formation should be supported, e.g. during the introductory seminars. In addition, a first date, e.g. for a meeting in the chat room, should be set at the beginning of the project work. The students suggested to have one of them act as a „team leader", coordinating the team. Regarding the use of media, it became clear that classic means of communication are important. Overall this evaluation of the first prototype showed that the INTEGRAL system is more than only the Internet based system.

## 4. REALIZATION

The following figures 3 and 4 show screen shots of the current realization of the INTEGRAL system.

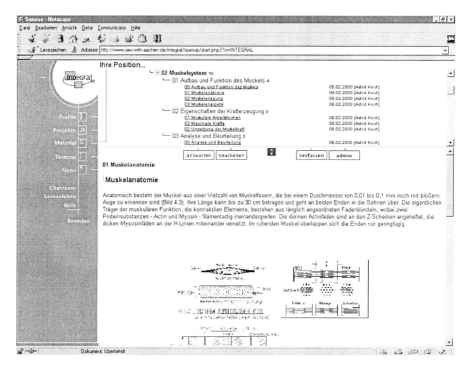

*Figure 3. Screenshot of the INTEGAL system (see text).*

After login to the system, the user can select what area of the system he or she wants to use. Figure 3 shows the basic layout: To the left is the main menu of the system. The upper part of the major part of the screen shows the different messages in the selected branch, the lower part gives a view of the selected message (here: study material about muscle anatomy). The menu has the following entries:
- **Profiles:** Information about the collaborating institutes (persons, information about courses etc.).
- **Projects:** Material for the student projects (e.g. task) and area were the students can store their own data.

- **Material:** User specific selection of teaching materials (e.g. text, video, case studies).
- **Dates:** Overview of dates.
- **News:** Overview of new messages (added after last login).
- **Chat room:** Possibility to participate in different chat forums or to initiate a new chat forum.
- **Bookmark:** Possibility to bookmark web addresses or INTEGRAL pages and to add short notes.
- **Help:** Standard help function.
- **End:** Return to login page.

Figure 4 shows an example of one of the specific modules. It is a VRML model of a VDT workstation and a checklist to evaluate it. This module was developed by the partners from the Institute for Occupational Physiology at the University of Dortmund.

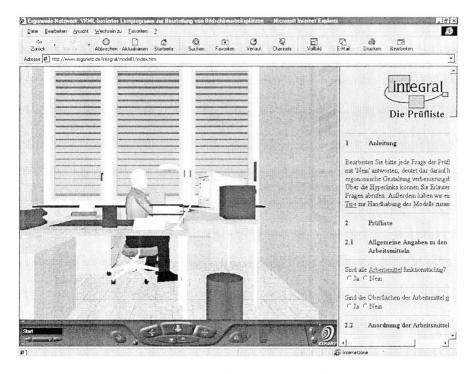

*Figure 4. Screenshot of the module "Evaluation of a VDT workstation"*

## 5. PERSPECTIVES

Currently the INTEGRAL system is used to support the "local" courses at the participating universities. At the time of the writing of this paper, new funding is about to be raised that will allow to extent the scope of the INTEGRAL system. Twelve additional partners from German universities will join the project group. INTEGRAL II, as this second stage of the project will be called, will considerably broaden the scope of the ergonomic content and the number of users.

## 6. ACKNOWLEDGEMENTS

The INTEGRAL project was funded by the Universitätsverbund Multimedia in North Rhine-Westphalia. Partners in the project are the University Essen, Fachgebiet Ergonomics in Design (R. Bruder, A. Kalweit & A. Pankonin), the University Siegen, Ergonomics (H. Irle, E. Keller & H. Strasser), the Institute for Occupational Physiology at the University of Dortmund (D. Gude & W. Laurig), and, as coordinating partner, the RWTH Aachen, Institute for Industrial Engineering and Ergonomics (D. Gude, H. Luczak, M. Rötting & S. Schneider). The homepage of the project can be found at http://www.iaw.rwth-aachen.de/integral/.

## 7. REFERENCES

Beitinger, G. & Mandl, H. (1992). Entwicklung und Konzeption eines Medienbausteins zur Förderung des selbstgesteuerten Lernens im Rahmen der betrieblichen Weiterbildung. In: Deutsches Institut für Fernstudienforschung an der Universität Tübingen (ed.). Fernstudien und Weiterbildung (pp. 92-126). Tübingen: Deutsches Institut für Fernstudien.

Fischer, M. & Hägebarth, F. (1999). Die technologischen Voraussetzungen zum Lernen und Lehren in der Informationsgesellschaft. In: U. Beck & W. Sommer (eds.). Learn-Tec 99 – 7. Europäischer Kongreß und Fachmesse für Bildungs- und Informationstechnologie. Tagungsband (pp. 597-605). Karlsruhe: Karlsruher Kongreß- und Ausstellungs-GmbH.

Geyer, W., Eckert, A. & Effelsberg, W. (1998). Multimedia in der Hochschullehre – Tele-Teaching an den Universitäten Mannheim und Heidelberg. In: F. Scheuermann, F. Schwab & H. Augenstein (eds.). Studieren und Weiterbilden mit Multimedia (pp. 170- 196). Nürnberg: BW Bildung und Wissen Verlag und Software GmbH.

Harms, I. (1998). Computer-vermittelte Kommunikation im pädagogischen Kontext. In: F. Scheuermann, F. Schwab & H. Augenstein (eds.). Studieren und Weiterbilden mit Multimedia (pp. 252-278). Nürnberg: BW Bildung und Wissen Verlag und Software GmbH.

Scheuermann, F. (1998). Informations- und Kommunikationstechnologien in der Hochschullehre – Stand und Problematik des Einsatzes computergestützter Lernumgebungen. In: F. Scheuermann, F. Schwab & H. Augenstein (eds.). Studieren und Weiterbilden mit Multimedia (pp. 18-53). Nürnberg: BW Bildung und Wissen Verlag und Software GmbH.

Schwuchow, K. (1997). Wissenstransfer mit neuen Informations- und Kommunikationsmedien. Personal 11/1997, pp. 552-557.

Stockfisch, J. & Sigel, U. (1997). Interaktives Lernen in der betrieblichen Praxis. In: K. Schwuchow & J. Gutmann (eds.). Jahrbuch Weiterbildung 1997 (S. 90-93). Düsseldorf.

Unz, D. (1998). Didaktisches Design für Lernprogramme in der Wissenschaftlichen Weiterbildung. In: F. Scheuermann, F. Schwab & H. Augenstein (eds.). Studieren und Weiterbilden mit Multimedia (pp. 308-334). Nürnberg: BW Bildung und Wissen Verlag und Software GmbH.

*Matthias Rötting*
*Aachen University of Technology*
*Aachen, Germany*

V. SINHA , B. H. WALKE

# AN INTERNATIONAL COLLABORATIVE NETWORKED VENTURE FOR MOBILE COMMUNICATION STUDIES

**Abstract.** In today's world free flow of technology, trade, commerce and capital are integrating the global economy bringing societal upliftment all around. It is essential to have trained manpower with international bias who would be responsible for this global integration. We present an international linkage scheme in one of the vital areas of transportation: communication. Mobile communication is playing a crucial role in shaping the lives of people, how they do commerce or even governance. The technology of mobile communication is dynamic and will not mature at least for foreseeable future. Since the development of mobile communication is global, there is an urgent need for creation of such an international study centre offering highest possible expertise and also acting as the warehouse of information flow on new developments. The concept, which uses multimedia tools and web-based educational set-up, is of a virtual global university with Institutions in different continents jointly involved in pedagogy and research. The manpower so trained will be conversant at least with two different work cultures and ethos of two distinct societies.

## 1. INTRODUCTION

Communication as a vehicle for the socio-economic progress of a society is today axiomatically accepted as the fourth dimension of transportation after land, sea and air transportation; as communication is intimately linked with prosperity and social order. The new millennium gets heralded with the technological solutions for global connectivity. Telephony networks (PSTN etc.) along with terrestrial cellular systems augmented by mobile satellite communication systems make it possible to have information transfer between any two points on the earth. Same applies to Internet services. The present decade will see emergence of mobile communication as the main technological platform for multitudes of communication and multimedia applications just like the telephones served for basic communication connectivity till the last decade. Present trend of merger of audio, video and data has introduced a paradigm shift in what was hitherto known as communication. Mobile communication will play a crucial role in shaping the lives of people, how they do commerce or even governance. The technology of mobile communication is not static and will not mature at least for foreseeable future. The third generation mobile systems (UMTS, Universal Mobile Telecommunications System/IMT-2000, International Mobile Telecommunications-2000) paradigms are still evolving. Value added services based on UMTS are yet to be conceived. Since this development is global and not limited to a continent, there is an urgent need for creation of an International Centre for Mobile Communication Studies which will offer the highest possible expertise and also act as the warehouse of information flow on new developments.

The Centre should provide much needed and valuable focus on the subject not only for the traditional faculty-student participants from academia for their roles in creation, dissemination and application of knowledge in the field but also for corporate personnel, investors, executives, government, analysts, strategists, and marketing people besides the engineers. The objective of the Centre is to provide an environment where state-of-art research and development work will be carried out along with providing a platform for discussion and exchange of ideas for commercial exploitation of the recent developments in the field.

Though ushered in by leading enterprises communication technology, being a service-centered utility business, is driven by the consumer market. The telecom sector is getting deregulated almost in all the nations but the market environment can only be gauged by the local telecom companies because only the local people can understand the market that is influenced by the local conditions. Focussed attention is to be paid to facilitate expansion of such markets. There are vast potentials in the sector in the developing world, particularly in Asia and Latin America. It is for this reason that the Centre should have established branches in these continents. And it is for these cross-fertilization reasons that the manpower to be trained must have an international bias before they graduate to take up meaningful positions in the telecom field.

The Centre will consist of affiliating institutions across the continents interlinked and networked in the common pursuit outlined above. The linkages amongst the institutions necessarily have to be loose and still form a cohesive entity to really make the globe a common communication village.

## 2. OBJECTIVES

The purpose is to establish a network of academic institutions, in different continents, having excellent credentials in the field to meet the following with cross-continental bias:

(i) Capacity building in the field through training engineers and through young researchers;
(ii) Interaction with academia and industries through networking and twinning arrangements;
(iii) Generate environment for economic and social developments in the region and thereby become a role model for such interactions in the developing world.

The programmes that will meet the above objectives are

(i) *Research programmes*: Research activities, in the broad area of mobile communication, including
    (a) Wideband systems; networking and switching
    (b) Wireless local area networks and wireless local loop systems
    (c) Third generation systems
    (d) Internet platform for multimedia applications including e-commerce
    (e) Network security

(f) Digital Mobile Systems
(g) Mobile satellite communication
(h) Teletraffic studies
(i) Evolving market studies
(j) Teleteaching

will be persued. The topics are illustrative and not exhaustive. Expertise would already exist in most of the Institutions in several of the above areas. Efforts will be made to recruit young researchers to complement missing areas.

(ii) *Teaching programmes*: In order to sustain (i), present academic programmes will be suitably augmented to provide specialization in Mobile communication Studies; i.e., a new stream of specialization will be generated. Successful candidates will be awarded M.Tech /Ph.D. degrees according to the regulations of the Institute. A shorter 'Diploma in Mobile Communication studies' programme for participants from serving industries, utilities etc. will be generated.

(iii) *Non-formal educational programme*: Continuing education for enhancement of knowledge to the working engineers from industries, research institutions, service providers etc. as well as information dissemination programmes for executives of these sectors besides government and regulatory agencies are necessarily to be part of the activities of the Centre.

(iv) *Consultancy Programme*: The faculty and research staff will be encouraged to provide consultancy to agencies, both governmental and non-governmental. This could also be in the form of commercial exploitation (e.g., technology transfer) of work carried out under (i), through development of region specific prototypes or particular software, say for an industry. It may also be related to the application of technology for societal needs through market research.

(v) *International linkages programme* : The emphasis is to increase the linkages through exchange programmes of students and faculty, international delivery of education, organization of conventions at periodic intervals.

## 2.1. Research Programmes

The Institutions shall engage themselves in the state-of-art activities in all or selected areas depending on own expertise. The faculty and the scientific staff will be recruited for these activities or seconded to the programme by the respective organizations. It is imperative to have industrial bias in the research activities.

## 2.2. Teaching Programmes

The Centre will be engaged in manpower training through postgraduate programmes only. It is the teaching programme under this proposal which is different from the usual teaching activities of the Universities. The programmes are designed to impart state-of-art knowledge, in techniques, tools and skills, in the field of communication engineering with emphasis on mobile communication. In addition, they will imbibe multinational culture of problem solving. Such graduates would be needed in a wide range of application areas, including industrial and nonindustrial research, planning, development, implementation, fabrication, distribution and maintenance of various mobile communication systems and services.

*2.2.1. Master's Programme*

It is anticipated that the main thrust of manpower training will be through the Master's programme specializing in Mobile Communication Studies. Students registering for the programme shall pursue the programme atleast at two of the participating Institutions (in two different countries), including the host institution from where the student will get his degree.

The Institutions will devise their curriculum such that the coursework, to be taken by a student, gets completed in the first two semesters. These could be through courses offered by the instructors of the respective institutions or offered, through teleteaching, by instructors located at different institutions or a mix of the two. In case of teleteaching, it is anticipated that the Institutions will accept the grades awarded by the remote tutors as equivalent to the grades of the courses the Institutions offer. There could be a possibility of an Institution insisting on imparting compulsory courses in the traditional manner (and not allowing teleteaching of the compulsory courses). At the end of course work the student will embark upon the project work towards his master's thesis under joint supervision of faculty, one at the host institution and one at a different location, the two faculty members having decided about the project a priori through mutual consultations. The student will move over to the other Institution to carry out the project work. The concept is, the courses are primarily done at the host institution but the project work at the other Institution. At the end of the Project work, normally six months, the student returns back to the host institution and submits the work carried out, as above, for the award of the Master's degree. This period will also be utilized by the student to learn the culture of the other place. Typical compulsory courses would include four subjects on Queueing Theory, Telecommunication Networks, Digital Mobile Radio Networks and Communication Theory which would have three hours of lectures per week for a semester of 14-16 weeks duration. Elective courses may include subjects like Satellite Communication, Information and Coding Theory, Data Networks, Stochastic Simulation, Digital Signal Processing, Digital Switching, VLSI Design, Speech Processing, etc. A student is expected to take four of these subjects, besides the compulsory subjects, towards fulfillment of his course-credit requirements.

*2.2.2. Ph.D. Programme*

A student registered for programme leading to the Ph.D. degree will spend at least one year away from the host Institution for work preparatory to the Ph.D. thesis. He will earn his grade while he is away according to the rules of the Institution he is visiting. This grade will be accepted by the host Institution.

*2.2.3. Diploma Programme*

A Diploma programme of one year (two semesters and a short project work) will be designed for specific industry participants through course work, including teleteaching.

*2.3. Non-Formal Educational Programmes*

Perhaps the most important activity of the Centre and the associated Institutions will be related to the Non-Formal Educational Programmes. The technology is expanding so rapidly, and value added services especially multimedia communications systems products are being added so frequently that there is a need to continually update and disseminate information regarding the field. Aspects of mobile communication are not appreciated by large cross-section of population, including policy makers, bureaucrats, government bodies, service providers etc. Even the executives of communication sector enterprises as well as regulatory authorities require continual upgrading of their awareness of the trends in the communication technology and services. The Institutions will be engaged in dissemination of information through sustained non-formal educational programmes. These programmes could be run by the Institutions utilizing the best expertise available on its faculty across the globe in telelearning mode (remote tutoring). Besides remote-tutoring, in-house continuing education programmes or programmes at the site of large companies/organizations will also be organized at regular intervals. A joint continuing education programme can also be held in third countries. The Institutions will also develop broadband interactive educational packages in the area for its utility by consumers. It is anticipated that the non-formal educational programmes would not only be self-sufficient but also to large extent subsidize the other programmes.

*2.4. Consultancy Programme*

Each Institution will provide its expertise to agencies (government/non-government/industry) according to the norms of the Institution. It may also act as liaison between a client and an expert at another Institution. The Institutions may develop appropriate software for mobile communication. Communication being a customer-driven utility industry, there could be a regional bias, especially in developing countries, so far as the telecom market is concerned, depending on the civilization, language and ethos of people. The market research or even application or transfer of technology will, therefore, be region dependent. The Institutions may take up commercially exploitable development work with the participating industries or agencies. Basic issue is, because of the expertise available in an

Institute, the industries would look towards the Institution for providing cost-effective solutions for services and technology adaptations.

*2.5. International Linkages Programmes*

The emphasis on creation of the Centre is on international linkages. Curriculum, mode of delivery of the curriculum, expertise to be utilized for the same will be jointly decided by the Institutions. The linkages would be in the form of students exchange programmes and faculty exchange programmes. Impetus on these two for global impact of mobile communications is essential. Faculty will be encouraged to spend teaching a semester every two years at another Institute as part of their work load.

It is essential that there is a close interaction among the Institutions of the Centre. International co-operation flourishes only when personal contacts find deep roots. Towards this, the Centre will organize a yearly convention on topics of current interest, experts addressing on the subject. The Convention would also be open to delegates, say from industry, for a fee.

## 3. ADMINISTRATIVE STRUCTURE

The Centre will have a Governing Council headed by a Professor of one of the Institutions with its Secretariat initially at one of the Institutions responsible for initiating the scheme. The Head of the Centre will have a tenure of three years with the possibility of re-election by the members for a second term. The members of the Council will be a Professor from each of the participating Institutions together with one nominee each of the funding organizations.

The role of the Council is to superintendent functioning of the Centre, especially the international linkage programmes, generate resources from various funding agencies including the industries and will be answerable to the funding agencies, and to develop and suggest details of academic programmes for implementation by the Institutions. For academic programmes the Institutions will be under the academic control of the respective Universities of the participating Institutions.

The Council will formally meet at least once a year at one of the Institutions.

For their efforts towards the activities of the Centre each of the Professors on the Council will be financially compensated according to the norms drawn by the Council.

## 4. REQUIREMENTS FOR ESTABLISHMENT OF THE CENTRE

For the proposal to take off, following initial requirements are projected:
- (i) At least three Institutions;
- (ii) Guaranteed funding for five years; basically towards salary for faculty and staff; scholarship for students while they would be in different country (on lines with DAAD scheme); funding to meet the travel costs;

infrastructure development; laboratory tools and workstations; digital library, etc.
(iii) MoUs amongst the Institutions; MoUs with the funding agencies.

Discussions between the RWTH, Aachen (COMNETS), Aalborg University, Denmark and Indian Institute of Technology Kanpur reveal that there exists considerable interest in the scheme. Informal discussions with some of the industries have also revealed their enthusiasm for the programme.

## 5. CONCLUSION

An attempt has been made to project a blue print for establishment of an international collaborative networked venture which may form the basis of a true global university. The specific area from where this concept can take roots is the area of mobile communication studies. Such a scheme will produce trained manpower with cross-cultural exposures, an essential parameter for business success in the present situations.

*Vishwanath Sinha*
*Indian Institute of Technology Kanpur*
*Department of Electrical Engineering*
*Kanpur, India*

*Bernhard H. Walke*
*Aachen University of Technology*
*COMNETS*
*Aachen, Germany*

V. SINHA

# A VIRTUAL CLASSROOM AND MORE FOR GLOBAL EDUCATION

**Abstract.** Using the present-day multimedia tools it is feasible to create a virtual classroom over the Internet as a possible alternative to conventional education. In the conventional system a student enrolls for a certain number of courses, attends lectures, asks questions, listens to his fellow students asking questions, submits assignments and appears at examinations to earn credits. We have developed a system over Microsoft Windows platform keeping an eye on portability to Unix/Linux in future. It provides a truly interactive live session between the teacher and students. The interaction is through text, viewgraphs, graphics, free-hand drawing/writing, voice, audio and video. A database server with SQL-database manager provides web-based administrative chores like registration, browsing of lecture materials and downloading of pre-recorded lectures etc. Lectures are given from a teacher's console and received on students' consoles and mimic the classroom scenario in terms of interactions between the teacher and taught. Besides the conventional applications as mentioned above, the system could also be effectively used for interaction between experts and engineers located at different sites world over on subjects including on economic practices in a particular country.

## 1. INTRODUCTION

Present day technological tools enable us to mimic the actual pedagogic process. A student, desirous of enrolling himself in an academic programme of an Institute

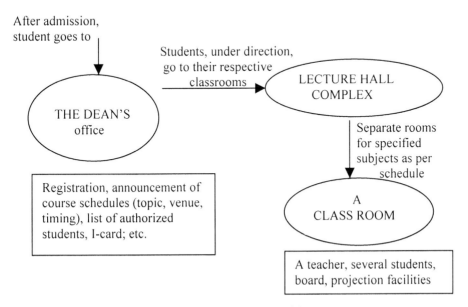

Figure 1. A Simplistic Model.

submits his credentials to the Institute. He is enrolled as a student if he fulfils the criteria of admission. On admission, he follows the prescribed guidelines for choosing courses and starts attending classes. The situation, in a simplistic manner is shown in Figure 1. With the availability of reasonably cheap servers, computing facilities and a fairly reliable telecom and data network, it is possible to mimic the scenario, thus generating the concept of a "Virtual Institute". The model developed is shown in Figure 2. Realizing the potential of multimedia for teaching, we at our Institute, under the Telematics Project, have developed a Remote Tutor product providing user-friendly multimedia tele-teaching environment. We describe its

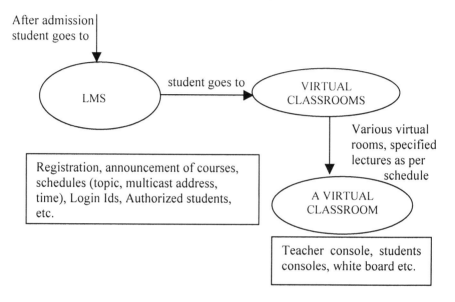

*Figure 2. Remote Tutor Concept : Realizing Virtual Institute.*

features in this paper. It has been tested and is under field trial at the moment. It is a complete technological solution to Internet based education and is aimed at facilitating the education system in sister institutions by providing a solution for implementing a network based classroom. The students need not be physically present in an institution to attend the classes or appear at the examinations. So is true of the faculty members responsible for delivering the lectures.

## 2. REMOTE TUTOR

The complete remote tutor package consists of two parts: a database server with ASP based implementation for providing administrative services; and another server, Lecture Manager Server (LMS), having multimedia capabilities for implementing the virtual classroom over the Internet. The application requires underlying TCP-IP connectivity with multicast support. Figure 3 describes the interconnection between Remote Tutor application and LMS. The present implementation of LMS is on Windows NT based Server with SQL-database manager. It provides web-based interface for Registration, Browsing of Lecture materials and downloading of pre-recorded lectures, for Announcement and for other administrative purposes, etc. The ftp interface of the LMS server allows automatic/manual downloading and uploading of lecture related material from/to the lecture database of LMS, by the Remote Tutor Application. A student/ Instructor registers with the LMS server by submitting the respective registration

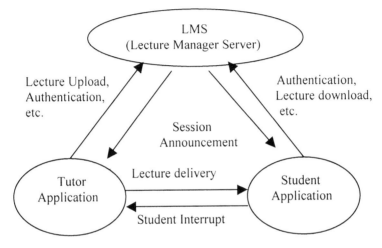

*Figure 3. Connectivity diagram.*

form on the web. The administrator of the LMS server checks the forms and once the necessary formalities are over, the form is sent to the database. A login-id and password is sent by mail to the student/instructor. This provides access to all the services of the LMS server for which the student/instructor is authorised. The functions of LMS are primarily of two kinds: administrative control and management; and other functions. The administrative control and management functions consist of (i) course announcement and maintenance, (ii) lecture scheduling, (iii) student and instructor registration/authentication, (iv) central repository for course lecture material, (v) announcements, (vi) browser based access to lecture material and recorded lectures, (vii) database management and backup, (viii) maintenance of virtual classroom over Internet and (ix) control of live sessions, etc. Other functions include (i) assignment download and submission, (ii) support for chat session amongst the students, (iii) offline student instructor interaction through email/chat support, (iv) support for conducting on-line examinations, and (v) providing tools for content generation, etc. Figure 4 gives the block level description of the LMS server and its interface with the outside world.

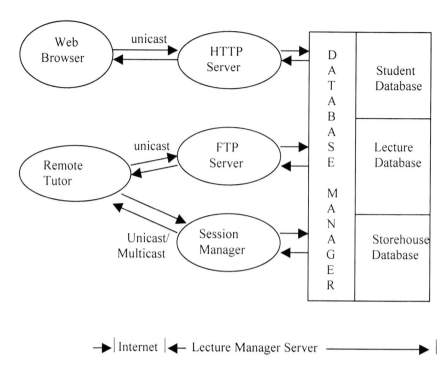

*Figure 4. Lecture Manager Module and its interface.*

## 3. REMOTE TUTOR APPLICATION

The multimedia teleseminaring tool, developed over Windows platform, is an application that simulates the virtual classroom environment over the Internet. It provides an interactive live session between the teacher and students. The interaction is through text, viewgraphs, graphics, free-hand drawing/writing, voice, audio and video. Though this component has been designed to operate in conjunction with the LMS server, it can also be used in stand-alone mode for delivery of a seminar over Internet. It uses COM based interface for accessing objects prepared using other applications, e.g. slides of different types (.doc, .ppt etc.), audio and video clips etc. It uses DirectShow standard interface for audio and video capture.

*3.1. Features of Remote Tutor Application*

The features of the application are given below.
- Lectures are given from a teacher's console and are received on students' consoles located on the same LAN/ on a remote LAN connected over a WAN.
- Teacher can deliver a recorded lecture or present the lecture on-line or mixture of the two can be used.

- An interactive live tool is available between the teacher and the students.
- Students can ask questions after seeking permission from the teacher. The student may ask for a discussion by interrupting the teacher. The interrupt message pops up on the teacher's console along with the student's identity. Teacher may postpone, cancel or acknowledge the interrupt. The student receives the control if the teacher acknowledges. The student can multicast his query and return the control when over. The teacher, however, can take back the control at any time.
- Interaction between the teacher and the students is through text, graphics, voice audio and video whereas students' queries are over audio channel.
- Students can join a lecture late and leave early.
- It provides support for preparation of lecture material, e.g., slides etc. and organise them in proper sequence for smooth delivery of lectures.
- Automatic uploading and downloading of lecture related material as well as online/offline recording of lecture are possible through this tool.
- It allows the student to switch on/off either or both the real-time streams, audio and video streams, depending on the available bandwidth at the student's end.

## 4. CONCLUSION

There is a need to develop tools for e-learning. We have explained the remote tutor tool that has been designed at our Institute to meet this requirement. It is a user friendly tool which works on the Windows platform. It is interactive. The teacher and the students can communicate with each other almost mimicking a real life classroom scenario.

*Vishwanath Sinha*
*Indian Institute of Technology Kanpur*
*Department of Electrical Engineering*
*Kanpur, India*

# PART VIII.

## SUCCESSFUL PRACTICE IN ENGINEERING EDUCATION -

### UNIVERSITY-INDUSTRY PARTNERSHIP, DESIGN PROJECTS

E. DOERRY, B. BERO, D. LARSON, J. HATFIELD

# NORTHERN ARIZONA UNIVERSITY'S DESIGN4PRACTICE SEQUENCE

*Interdisciplinary Training in Engineering Design for the Global Era*

**Abstract.** Introduction of computer technology, changing corporate structures, and global competition have significantly changed modern corporate design contents, placing increasing emphasis on individual problem-solving creativity, interdisciplinary collaboration, and teaming and project management skills. In this paper, we describe a program called Design4Practice that explicitly teaches these skills within a novel curriculum centered around a carefully crafted sequence of project-oriented courses. The motivations and goals of these courses are presented, as well as outcomes of program evaluation and a discussion of current efforts to extend the program to allow international teaming.

## 1. INTRODUCTION

The notion of "design" lies at the heart of engineering practice; engineering is fundamentally about creating novel solutions to modern technological problems. The complexity of the design enterprise has grown dramatically in recent decades, fueled by the ubiquitous introduction of computer technology, increased awareness of safety, social and environmental concerns, and a growing trend towards corporate globalization. As a result of these revolutionary changes, young engineers emerge into a workplace that is more challenging than ever before: they must be proficient with a variety of sophisticated computer-based design tools, be able to communicate effectively within a fast-moving electronic world, and be able to collaborate efficiently as members of multi-disciplinary, multi-cultural, globally-distributed design and development teams. In short, modern engineering practice places as much emphasis on individual problem solving, communication, and interdisciplinary teaming skills as it does on traditional technical engineering skills.

Despite these radical changes in the context of real-world design, university curriculums have been slow to adapt. Most programs still focus almost exclusively on teaching traditional technical skills (e.g., statics, digital logic, programming languages, etc.); students receive very little explicit training in how to *apply* these skills within modern corporate design and development contexts. As a result, engineering graduates enter the workplace with strong technical training, but ill-prepared for the creative, communicative, and administrative demands of modern engineering practice.

In an effort to remedy this deficiency, many educational institutions have begun to include some sort of "practicum" experience in their engineer design curriculums, in which students engage in a real-world design projects under the guidance of expert design engineers. Unfortunately, such courses are typically isolated pedagogical events, taught as just one more semester-long topic within a curriculum centered around traditional engineering courses. This approach fails to recognize that - perhaps more so than for any technical skill - the only way to fully develop creative, communication and teaming skills is through repeated exposure to design in a real world context: an ever-widening spiral of problem statement, design development, failure, and critical reflection.

In this paper, we describe a program that reverses traditional engineering curricula by situating technical engineering courses within a novel curricular framework centered around interdisciplinary, practice-oriented design training. This program, called Design4Practice, engages engineering students in a series of increasingly challenging courses in which student teams solve a wide variety of realistic design problems. In addition to traditional technical skills, heavy emphasis is placed on teamwork, communication, project management, and interdisciplinary coordination. In evaluating this six-year old program, we have found that graduates are better prepared to contribute immediately within modern heterogeneous corporate development teams upon leaving the university, and mature more rapidly to leadership roles within their disciplines.

In the following section, we provide a broad overview of the Design4Practice program and its curricular goals. In subsequent sections, we describe the component courses of the program in more detail. Closing sections discuss our efforts to evaluate the program, and speculate on future plans to internationalize the program to better prepare engineers for the global workplace.

## 2. THE DESIGN4PRACTICE PROGRAM

To address the challenges surrounding the teaching of real-world design skills, Northern Arizona University's College of Engineering and Technology (CET) has developed a four-year interdisciplinary sequence of classes called Design4Practice (Hatfield, Collier et al. 1995; Collier, Hatfield et al. 1996; Howell, Harrington et al. 1996; Larson 1999). This practice-oriented engineering curriculum, crafted with extensive input from industry, is designed to provide students with hands-on learning experiences and the continuous practice of a broad set of professional skills in order to better prepare them for careers as engineering practitioners.

Unique features of the program include:
- Cross-disciplinary collaboration in sequenced courses.
- Cooperative teaching and learning teams.
- The active participation of industry executives and engineers through teaching, program evaluation and project sponsorships.
- A required core for all engineering students incorporating the complete design cycle within industry simulated product development environments.

The overarching teaching objective for the Design4Practice program is the blending of theory and fundamentals with practice. Design process, methods and tools, and cognitive skills are integrated into guided problem-based activities in a just-in-time format – when the students need the technical and/or professional information, it is provided. In other words:

"Theory without practice is ineffective, ... and practice without guidance ... results in frustrations and failure to exploit the knowledge..."(Ulrich and Eppinger 2000).

The Design4Practice program objectives were developed in response to the call by industry for baccalaureate engineers to possess a broader set of skills in addition to their sound analytical and computer skills. This call was strengthen by our own observation of our student's experiences in senior capstone design during the late 1980's and early 1990's. These students, who had had no prior experience with true design, struggled with design process and problem solving, project management, and teaming issues. Their successes were limited; hampered not by a lack of technical knowledge, but because they lacked the set of skills that we call design.

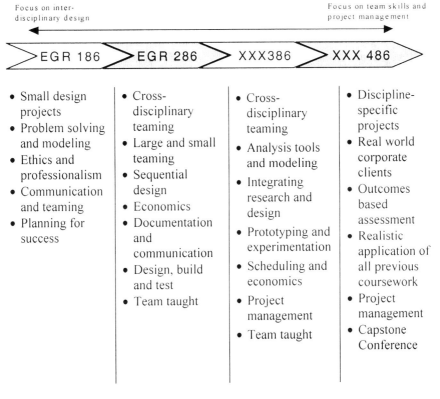

Figure 1. Overview of the four core Design4Practice courses, highlighting key features of each course. Each course also teaches discipline-specific technical skills associated with completing the project.

The strategy behind the Design4Practice sequence is straightforward: to introduce the students to the design process early in their college careers, and maintain a constant rate of increasing complexity throughout the four years of study. By graduation, the students are well versed in the design process and hence have a greater ability to successfully contribute in their first professional employment. The impact of the Design4Practice program on students has been evaluated since 1994. The Design4Practice program has been successful in reaching the objectives mentioned above, enhancing our students' ability to contribute and succeed in industry immediately upon graduation.

*2.1. An integrated curriculum*

The Design4Practice curriculum is built around four core design courses, one for each of the four years of the degree program: the freshman course, EGR 186 Engineering Design – Introduction; the sophomore course, EGR 286 Engineering Design - The Process; the junior course, xxx 386 Engineering Design - The Methods; and the senior course, xxx 486 Engineering Design - Senior Capstone[1]. Each course builds upon the previous course in the sequence by threading of topical content from course to course. The courses are team-taught by teams faculty and practitioners who are experienced in engineering design.

Each course in the sequence has its own mission, objectives, strategies and educational outcomes that map directly into this higher level program.

As indicated in Figure 1, the four-course curriculum begins with a focus on interdisciplinary design and teaming in generally; this focus gradually shifts towards development of discipline-specific teaming skills and project management as students mature and gain design experience. The following paragraphs briefly describe the courses in more detail:

*EGR 186: Engineering Design – Introduction*

This freshman engineering design course was developed in collaboration with colleges of engineering at the University of Arizona and Arizona State University and with the Arizona Community College system. Multiple, small, team-based engineering design experiences are emphasized along with problem solving techniques, teaming and research skills, oral and written communications skills, and computer-based tools for academic and professional success.

Multiple sections of EGR 186 are offered each semester with the number of students per section capped at thirty. By limiting the number of students per section, the section instructor can better monitor team processes during the many, small activities that are used to convey course content. During the semester, student teams may design projects such as: a weight bearing bridge constructed with paper, glue, thread, and pins; an electrical learning machine that indicates correct and incorrect responses to a number of multiple choice questions; and a gravitational potential

---

[1] The latter two courses, denoted xxx 386 and xxx 486, are discipline-specific. That is, the six disciplines within the CET (Civil, Environmental, Computer Science, Electrical and Mechanical Engineering) each teach their own sections (although 386 project teams remain cross-disciplinary). Thus, we use the notation xxx 386, for example, to represent CSE 386, ME 386, and so on.

energy-powered marshmallow launcher (the maximum launch distance was 168 feet). These projects motivate the students to do library and Internet research, to strive for innovative designs, and to experiment with alternative design solutions.

EGR 186 uses a limited form of team teaching where the section instructors meet weekly to ensure that each section is taught in a similar manner. These meetings include planning the course, developing and structuring team and student activities, and sharing insights.

*EGR 286: Engineering Design – The Process*

This three-credit sophomore design course introduces students to the engineering corporate environment through a semester-long, design-and-build robot project. The faculty play the role of division managers or the chief executive officers of an imaginary corporation; students play the role of employee engineers. Each student belongs to a multidisciplinary team of five to seven students and that team is one of three or four teams in a division/mega-team. In total, the virtual corporation has three engineering divisions with sixty to eighty engineers, three division managers, and one chief executive officer.

After being introduced to the company's culture, learning their roles in the organization, and being properly trained in teamwork, students are presented with an engineering problem by a customer. The remainder of the course is spent on requirements capture, specification writing, creativity and concept generation, design and analysis, documentation, building and testing a solution.

Care is taken to ensure that the students apply solid, rigorous project management techniques rather than an informal approach. Each team within a division is responsible for different parts of the project during different phases. The results of one phase are passed on to the next team to work on during the subsequent phase. The hand-off procedure reinforces the importance of clear and concise documentation. It requires good inter- and intra-team communications and it exposes the students to all aspects of the project.

The projects in EGR 286 are computer-controlled robotic devices with complex logic structures that interface with the "moving and grabbing" mechanisms via various sensors. Examples of projects executed in past years include robots to scoop and move contaminated soil, search and map small tunnels while retrieving artifacts, and play an adapted version of basketball.

The central goal of EGR 286 projects is to encourage creativity and cross-disciplinary tinkering. At the same time, however, the students learn to program and build real-time graphical user interfaces in Visual C++; use electrical and mechanical hardware; deal with subsystem interface issues; machine, assemble, and test; and create industry-standard proposal, design, and as-built documents.

*XXX 386 Engineering Design – The Methods*

XXX 386 emphasizes the use of analytical modeling skills and related computer tools; rigorous design methodologies with system level thinking about social, environmental, and cultural impacts; and sophisticated project and team management that includes careful planning, scheduling, cost estimation, and economics.

This course is again structured around a virtual company that consists of discipline specific divisions working on appropriate parts of a central project. The course organizational structure, depicted in Figure 2, allows the junior students to focus heavily on appropriate analysis and design methods for their discipline without losing the interdisciplinary nature of the course. Students within the course are generally divided into three to four project teams, all working on similar implementations of the same specification. As indicated in Figure 2, each project team includes several from each discipline; these subgroups collaborate as a cohesive project team to complete the project.

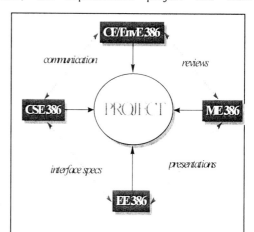

Figure 2: EGR 386 Project Team Structure: Each team consists of discipline-specific "subgroups" from each discipline; these subgroups typically are from 5-8 students in size. Total team size is thus around 25.

As with EGR 286, student feedback is elicited through a final postmortem report. In this report, the students are encouraged to reflect upon the course and to demonstrate their design process knowledge. The faculty team integrates the feedback, making course modifications when appropriate.

*XXX 486 Engineering Design - Senior Capstone*

XXX 486 is the final capstone course in the Design4Practice sequence in which students are expected to complete a real-world design project with little instructor intervention. Students in this course communicate directly and extensively with clients from outside the university; they apply knowledge and skills to solve a real problem, beginning with problem definition and finishing with a functional prototype.

Four separate and discipline-specific sections of XXX 486 are offered each spring semester. Within each section, the students are formed into teams to complete industry-sponsored projects. Such projects might involve, for example, design and implementation of a machine or component, an assistive device for the physically challenged, computer hardware or software, a land development or transportation plan, or a completed design for a structural system. Each team is supported by a project sponsor from industry (the client), a course instructor who acts as the project manager, and a technical advisor from the general faculty that has expertise related to the project. Projects are solicited from a broad spectrum of large and small corporations; some examples include Motorola, Honeywell, Mack Corporation, Boeing, and the U.S. Forest Service.

During the project cycle, design teams work with their sponsor, project manager, and advisor, to capture requirements; develop plans and schedules; analyze budgetary constraints; conduct requirements analysis; and design, implement, and test a solution. At the end of XXX 486, each student team is expected to deliver a complete and working product, prototype, or final design to their sponsor.

The curriculum culminates with the annual Capstone Design Conference, in which XXX 486 students present their projects to their industrial sponsors, technical advisors, project managers, NAU CET faculty, the general student body, many of the graduating student's parents, members of the College of Engineering Industrial Advisory Council (CEIC), and the general public. This conference follows the format of typical professional conferences, with students providing project overviews in a paper session; demonstrations of actual products are made during a subsequent poster and demonstration session.

*2.2. Lessons learned: assessment and improvement*

Since the inception of the Design4Practice program, we have continually modified and improved the program objectives and strategies. Until recently, this process was informal and loosely documented, and incorporated the input from three main constituency groups: students, faculty, and practicing engineers. Various types of student assessment techniques were employed as well as a solicitation of feedback from project sponsors and other interested industry participants. A summary overview assessment methods used is provided in Table 1. Of these methods, the student post mortem reports were the most useful. These individually written compositions provided us with a wealth of constructive feedback from the students on how well we were meeting our objectives and ways to improve the design course offerings.

In the Spring of 2000, we began implementing a more formal assessment and improvement process for the Design4Practice program. Briefly, each class will develop its own set of unique assessment methods that will may include a pre-semester and post-semester inventory of skills and knowledge that is coupled to a postmortem report. In addition to answering specific course questions via the postmortem, the students must analyze the changes they observed and recorded on their skills inventory.

*Table 1. A sample of* Design4Practice *program assessment activities*

| Strategy\Description | Purpose | Results or Status |
|---|---|---|
| Post Mortem Reports: Individual final reports. | • Assess student's understanding and application of the design process<br>• Obtain course feedback. | • Growth in process understanding noted between EGR 286 and EGR 386.<br>• Many minor and some major class changes made. |
| Holistic Writing: Pre- and post- examination of writing. | • Measure growth in writing skill over a semester. | • Growth in skill difficult to statistically measure over a one-semester period (Grueber, Larson et al. 1999). |
| Confidence with Problem Solving: A post-semester test on student's confidence in problem solving. | • Measures student's perception of their confidence and approach to solving problems.<br>• Examine freshman to senior design populations. | • Measurable improvement in confidence over the four years (Larson, Scott et al. 1998). |
| Student Skills Matrix: Pre- and post- student evaluation of their growth in various skills. | • Measures student's perception of growth in technical, process, thinking, communication, and teaming skills. | • Provides a reflective learning opportunity for students to objectively assess their perceptions and skill changes. |

This type of strategy effectively and simply quantifies outcome-related skills growth and knowledge accumulation, while documenting the individual's communication ability and providing an opportunity for honest self-reflection about learning.

## 3. FUTURE DEVELOPMENT: REMOTE PARTICIPATION AND INTERNATIONAL TEAMING IN DESIGN4PRACTICE

One of the most characteristic features of economic change in the last decade has been the growing trend towards globalization: through either mergers with foreign partners or expansion into foreign markets, many large companies have developed subsidiaries spread across national, cultural, and linguistic borders. As a result, design and development initiatives often involve teams or team members separated by time, distance, and culture. For example, an increasing common practice in software development is to spread development of a project between subsidiaries in, say, San Francisco and Frankfurt; the time difference between the two allows the project to move forward non-stop, 24 hours a day. Working in such international teams raises two distinct challenges:

*Coordination over distance.* Coordination of work within distributed teams is extremely challenging, particularly in highly interdependent projects (Dourish and Bellotti 1992; Rogers 1993). Success in distributed teams requires team members to become proficient with sophisticated software tools (known as *groupware*) and learn how best to apply them to support the particular collaborative needs of the team. For example, a team might rely on video-conferencing software to discuss project details, a shared drawing tool to jointly critique a design document, and a private group web page to archive shared resources.

*Cultural Differences.* A growing body of work indicates that cultural and social differences within work groups (whether distributed or co-located) play an enormous role in determining team dynamics, i.e., whether or not a team "comes together" into a cohesive and productive team entity (Vick 1998). In many cases, minor cultural misunderstandings can create tremendous rifts in cross-cultural teams. For example, it is not unusual for European work schedules to be more flexible than those in the USA, e.g., a longer midday break followed by work later into the evening. Americans unaware of this difference might perceive continual noontime absence as "laziness"; team morale will suffer accordingly. Similarly, a lack of direct eye contact that is a sign of social respect in Asian countries is easily misunderstood as "shiftiness" by westerners. As a final example, differences in linguistic idioms, social codes, and modes of expression can easily lead to misinterpretation of the tone or intent of a communication between team members from different backgrounds. Social and cultural issues such as these may ultimately be far more important than technological issues in determining the success of cross-cultural teams.

In sum, participating effectively in the new global economy will require engineers to possess a new set of social skills and technological skills, in addition to discipline-specific technical skills. We believe international teaming will become increasingly common over the next decades; global corporations are placing increasing emphasis on international experience in hiring and promotion.

As part of a broad initiative to provide opportunities for international study and exposure to the CET curriculum, we are developing mechanisms to support international teaming within the Design4Practice program. Since the inception of the program, Design4Practice teams have consisted solely of co-located teams (i.e., resident NAU students). With recent advances in internet technologies, however, technical obstacles to participation of team members located elsewhere within the USA and the world have largely disappeared. We are currently developing or compiling an array of groupware applications — including video-conferencing tools, shared document viewers/annotators, tools for capturing and archiving design rationale and critical discussion, and tools for tracking the activities of distributed team members — to support remote participation in Design4Practice teams. These tools are being packaged into an integrated "virtual teamspace," a secure website that will serve as the hub of team development activities. With all team design activity firmly rooted in the virtual space, the geographic location of team members will become less relevant.

Despite these technical innovations, one major obstacle to international collaboration remains: the commitment of Design4Practice courses to the eventual *production* of the designed artifact. This means that, during the closing phases of the course, the emphasis of team work shifts from analysis and design to hands-on prototype development. The design teams simply must meet to physically test and explore various design prototypes. It is difficult to imagine how network technology could ever substitute for physical presence in this crucial experimental phase. Although local (i.e. as exchange students) participation is possible, asynchrony in semester timing between many European institutions and NAU makes this option impractical; students may have to miss *two* semesters at their home institution in order to take one Design4Practice course at NAU. In Germany, for example, fall semester starts in mid-October and runs through February; Spring semester starts in February and runs into July. In the USA, by contrast, fall semester begins in late August and ends in December; spring semester begins in January and runs through mid-May. Thus, a German student taking XXX 386 in the spring semester at NAU, would have to miss both fall and spring semesters at his or her home institution.

We are developing a scheme that addresses this obstacle while, at the same time, achieving our goal of exposing students to the concept of distributed design teams: international students will participate *remotely, via groupware technology* during the first part of a Design4Practice course, and then physically join the team in the closing weeks.

For example, Figure 3 illustrates how a German student might participate in the XXX 386 course offered every spring semester by CET. During the German fall semester, the student would participate in regular courses at his or her home institution, perhaps taking a somewhat lighter load in anticipation of remote participation in XXX 386 later in the semester. When the XXX 386 course begins with the NAU spring semester in January, the German student joins one of the design teams and begins participating in the early design process via the groupware tools available within the design team's website. This might include participating in on-line design meetings, evaluating and annotating design documents, and contributing analysis work. The student might also be sent prototype components to test in the lab at his or her local institution. When the German semester ends, the student joins the rest of the design team at NAU to participate directly in the critical closing phases of the project.

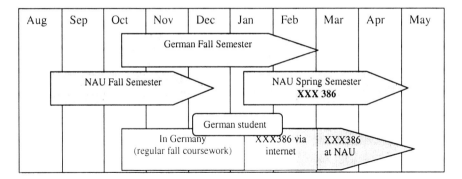

*Figure 3. Mixed-mode participation of a German student in CET XXX 386 course.*

The German student could also add one or two additional NAU courses via the same technology so that more than one course can be taken in the Spring semester.

By utilizing modern internet technology, international students are able to participate in Design4Practice courses without investing more than a single semester of their curriculum; the asynchrony in semester timing is actually used to an advantage.

## 4. CONCLUSION

Sweeping changes in corporate structure and the global economy have greatly altered the structure of modern corporate design and development teams and the expectations placed on their members. Project teams have become smaller and less structured; all team members are expected to be able to creatively solve problems and work both independently and as efficient collaborators within their working groups. In an increasingly global environment, teams are often separated geographically and may be cross-cultural in nature.

These changes in work context have increased the importance of project management, communication, and teaming skills in engineering graduates. The Design4Practice program developed within the CET at Northern Arizona University is designed to address these critical needs by situating traditional technical coursework within a curriculum centered around progressive, interdisciplinary sequence of project-oriented courses that mimic the structure and overall constraints of design in a real-world corporate environment. Our extensive and on-going evaluations of the program have shown that, in addition to exercising discipline-specific technical skills, these courses develop problem-solving and teaming skills that allow graduates to be immediately productive in the real-world design contexts.

One shortcoming of the Design4Practice sequence is that it does not expose students (both local and remote) to the concept and technologies of distributed teaming, or to working in cross-cultural teams. The international teaming initiative currently under development aims to address this need, by providing curricular and technological frameworks to allow international students to more easily participate in Design4Practice courses.

## 5. ACKNOWLEDMENTS

The CET Design4Practice program is the result of the collaboration of a large group of dedicated faculty. Other faculty who have played a central role in creating this program include M. Neville, D. Hartman, T. Baxter, S. Brinkerhoff, E. Brauer, C. Dryden, P. Eibeck, S. Nix, W. Odem, E. Penado, and D. Scott.

## 6. REFERENCES

Collier, K., J. Hatfield, S. Howell and D. Larson (1996). A Multi-disciplinary Model for Teaching the Engineering Product Realization Process. *1996 Frontiers in Education Conference*, Salt Lake City, UT.

Dourish, P. and V. Bellotti (1992). Awareness and Coordination in Shared Workspaces. *Proceedings of ACM CSCW'92 Conference on Computer-Supported Cooperative Work*, Toronto, Canada.

Grueber, S., D. Larson, M. Neville and D. Scott (1999). "Writing 4 Practice in Engineering Courses: Implementation and Assessment approaches." Technical Communications Quarterly 4th Quarter.

Hatfield, J., K. Collier, S. Howell, D. Larson and G. Thomas (1995). Corporate Structure in the Classroom: A Model for Teaching Engineering Design. *Proceedings of 1995 Frontiers in Education Conference*, Atlanta, GA.

Howell, S., T. Harrington, D. Larson, K. Collier and J. Hatfield (1996). A Virtual Corporation: An Interdisciplinary and Collaborative Undergraduate Design Experience. *1996 Design for Manufacturability Conference*, Irvine, CA, ASME.

Larson, D. (1999). A New Role for Engineering Educators: Managing for Team Success. *MRS Spring 2000 Conference Proceedings*, San Francisco, CA.

Larson, D., D. Scott, M. Neville and B. Knodel (1998). Measuring Student's Confidence with Engineering Problem Solving. *1998 ASEE Conference Proceedings*, Seattle, WA.

Rogers, Y. (1993). "Coordinating Computer-mediated work." Computer Supported Cooperative Work: The Journal of Collaborative Computing 1(4): 295-315.

Ulrich, K. T. and S. D. Eppinger (2000). Product Design and Development. Boston, Irwin/McGraw-Hill.

Vick, R. (1998). "Perspectives on and Problems with Computer-Mediated Teamwork: Current Groupware Issues and Assumptions." ACM SIGDOC *Journal of Computer Documentation 22(2): 3-22.

*Eckehard Doerry, Bridget Bero, Debra Larson, Jerry Hatfield*
*Northern Arizona University*
*Flagstaff, AZ, USA*

M. ROHS, W. MATTAUCH, J. CAUMANNS

# FURTHER PROFESSIONAL EDUCATION BASED ON WORKING PROCESSES[1]

*Workflow Embedded Training in the IT-Sector*

**Abstract.** There is currently an acute shortage of qualified personnel in the information and telecommunications sector. In order to solve this problem over the medium-term, the German Federal Ministry for Education and Research (*Ger: BMBF*) has ordered the set-up of a Regulation Procedure of the German Federal Institute for Vocational Training (*Ger:BiBB*) for (re)structuring the further education system in the ICT. Parallel to the Regulation Procedure and as part of the "New Media in Education" program, the BMBF, in collaboration with industrial partners and educational institutes, is promoting a project being carried out by the Fraunhofer Gesellschaft under the overall control of the Fraunhofer Institute for Software and Systems Engineering (ISST). The goal of the "Workflow-Embedded Training" (APO) project is to develop and elaborate an innovative system for further IT training.

## 1. INTRODUCTION

There is currently an acute shortage of qualified personnel in the information and telecommunications sector. According to economic experts, there will be a lack of 723,000 specialists in the field of information and telecommunication technology in Germany by the year 2003 (Bitkom 2001). However, more than 30 000 IT-specialists and 60 000 engineers are unemployed. One reason for that dilemma is that there is no real system for IT-qualification apart from university. To cope with the rapid innovation cycles improvement of advanced vocational training is an important task.

In order to solve this problem over the medium-term, the German Federal Ministry for Education and Research (*Ger: BMB+F*) has ordered the set-up of a Regulation Procedure of the German Federal Institute for Vocational Training (*Ger:BiBB*) to (re)structuring further education. In this way, further qualification in the IT field is to be made available to people with previous job experience in the IT-sector or those switching into the field , college graduates and persons with job experience but without the appropriate degree. Next to the Regulation Procedure and as part of the "New Media in Education" program, the BMBF, in collaboration with social partners, is promoting a project being carried out by the Fraunhofer Institute for Software and Systems Engineering (ISST), named "Further Professional Education based on Working Processes" (APO) (figure 1). This project aims to organizing vocational training work-integrated and based on business processes is to be

---

[1] This research was supported by the German ministry of research and technology (bmb+f) as part of the research project "APO".

developed in collaboration with industry partners and educational institutes to promote further professional education in the IT sector.

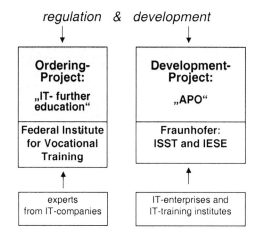

Figure 1. Parallelism of regulation and development.

## 2. THE NEW GERMAN STRUCTURE OF FURTHER VOCATIONAL TRAINING IN THE IT SECTOR

As a first result of the Regulation a structure was developed. In the structure about 30 professional profiles on three levels are defined: Specialists, operative and strategic professionals.

The lowest level are IT specialists. The prerequisite of this level are the skills of the IT-related vocational training. The specialists are subdivided into the fields:

- Analysts, e.g. IT System Analyst
- Developers, e.g. Software Developer
- Coordinators, e.g. IT Project Coordinator
- Administrators, e.g. Network Administrator
- Advisers, e.g. IT Business Systems Adviser
- Others, e.g. IT Supporter

On next higher level, operative professionals are defined: IT-Engineer, IT Manager, IT Consultant and IT Commercial. Prerequisite of these profiles degrees on the specialists level. Staff on this level are responsible for the operative businesses of a company.

Skills on the level of operative professionals are constitutive for a training to strategic professionals, the highest level of the new structure. They are qualified for the conduction of a company or a department. On this level are two profiles defined: one profile with business orientation (It Business Engineer) and one profile with technical orientation (IT Systems Engineer).

Degrees on the level of the operative professionals are thought to be comparable with a Bachelor degree, strategic professionals have a qualification similar to a Master degree (figure 2).

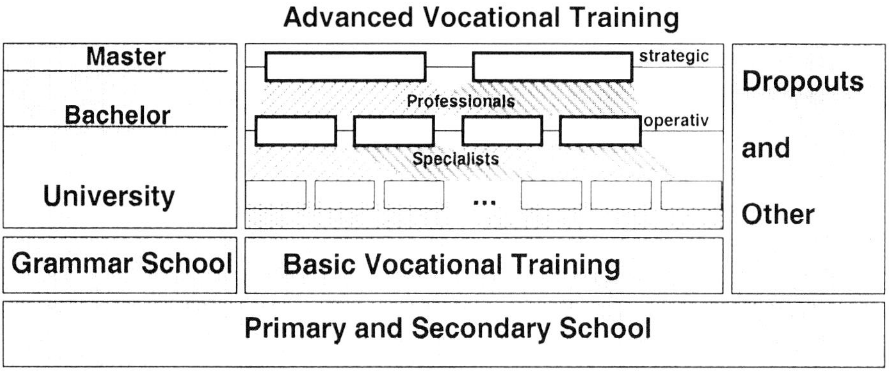

*Figure 2. New System for Advanced Vocational Training in the IT Sector.*

## 3. DIDACTIC CONCEPT

Fraunhofer ISST developed a basic concept to realize the structure described above. In collaboration with IT-experts (Deutsche Telekom AG, Deutsche Post AG a. o.), quality standards will be conceived which are aligned to the requirements of the IT branch. Graduates from IT basic vocational training courses, university dropouts and unemployed IT-specialist are the target groups. They will be offered new carrier prospects by means of further education (Grunwald & Rohs 2000).

*3.1. Requirements*

The basis for curricular decisions are the requirements of an alliance of representatives of employers and unions (http://www.iid.de/schule/it-fortbildung/markierungspunkte.html). According to their list of recommendations the advanced vocational training for IT experts should be:
- referring to the IT-related basic vocational training
- matching horizontal and vertical development of competencies
- permitting the participation experienced practitioners
- allowed the collaboration of companies and professional training providers
- implementing training measures in projects
- building curricular decisions on references

In addition to these political recommendations, the basic didactical concepts result as well from the requirements of the IT-sector.

So the IT-specialists must cope with changed objectives and different working environments. They work in interdisciplinary teams and increasingly customer-oriented. Moreover, IT-specialists are urgently-needed experts who can only rarely be taken out of the production process. Therefore it is necessary for them to:
- improve their professional skills
- strengthen their social capabilities
- develop auto-didactic capabilities

Advanced vocational training must take place within the working process through learning by doing. Thus acquired knowledge is highly product- and project-oriented. These informal learning strategies are common in many working processes and they comply well with the demands of the IT branch (Marsick & Volpert 1999). We intend to integrate them in a educational concept. Furthermore, this basic concept aims at developing self-learning strategies and impart the fundamentals of information technology. With these capabilities a IT specialist is able to keep up with the technical progress. As a result the educational concept has to be
- embedded in the companies' workflow,
- oriented on the practice,
- adapted on the individual skills, and
- oriented on competencies.

*3.2. Workflow Orientation*

The goal of the APO Project is to embed vocational training within the workflow and to consider individual learning efforts as part of a learning organisation (Dehnbostel 00; Albrecht 99). IT technicians will be trained on-the-job and on-demand in a project-like manner.

Within the working process, learning is always stimulated when employees are confronted with new or difficult tasks. But those informal learning processes occur unsystematically and without any outward signs (Livingstone99; Kirchhöfer00). In contrary to this, training courses are more structured and transparent, but they do not refer very well to the tasks and demands of the processes.

*3.3. Reference Projects*

A workflow-embedded qualification requires must be related to learning targets that are based on practical experience. For this reason, the first element of our concept are real-work assignment. But these practice projects are very different, so that there is no quality standard for the qualification as a whole.

That's why the key element of our concept are so-called *reference projects*. They represent typical and ideal business processes of a professional profile. Together with well-known IT-enterprises we search for suitable projects. Projects are suitable when they are up-to-date, documented, complete, typical and instructive. It is also possible to combine parts of different projects, e.g. the processes "change management" and "performance management" for a network administrator. The universal elements of the processes will be identified and merged into a so-called reference project. So the reference project is not a case study, but a generic scheme and a basis for a workflow embedded training. For example, the basis of the reference project for a software developer was the development of a e-commerce system carried out by *Software AG* (Tübingen).

The reference project has at least three functions:
- It enables the assignment of knowledge to a certain professional field,
- serves as a model for the ideal application of a tool or a method and the realization of own projects and
- serves as the quality standard for the qualification.

The reference project has the task to close the gap in the workflow between the work assignment and the requirements of advanced vocational training. This allows the adaptation of the individual work assignment in a way that all participants of the IT-qualification fulfil the same tasks.

## 4. IMPLEMENTATION

The strong focus on work and learning assignments requires a methodology of instruction and assistance., that is suitable for the different general conditions in companies ( e.g. the learning culture, size of an enterprise, number of participants, coaching personal a.o.).

### 4.1. Learning Arrangement

The arrangement of the implementation is oriented on the decentralised human resources development approach (Staudt & Kriegesmann 1999). Superiors are in a central capacity . They are interested in well qualified staff, know about individual competencies of their staff and are able to provide the conditions for learning of the workplace. However, in most cases superiors aren't able to manage vocational training by themselves. Participants needs help on a special questions and support to organize their learning process. That's why various persons such as specialists, mentors and trainers should support the learner.

Thus resources of an enterprise can be used better to improve human resource development. Therefore the APO concept is not only a contribution to a learning organisation but also a support and an incentive.

In case there is not enough qualified staff to provide sufficient feedback to the participant external specialists can attend to these part(s). Small enterprises could build (specific) networks to support one another. In this way one trainer or one specialist can instruct participants of different enterprises. Special internet tools can support communication and collaboration.

## 4.2. *Informal and self-oriented Learning*

The aim is to coach and accompany the participant in his learning processes rather than to teach him detailed information. Many participants are highly experienced but have not yet acquired a certificate of their qualification. The practical skills and the reflection of their experience can be a basis for the instruction of theoretical complexes. That's why a iteration of experience and reflection for the learning process are fundamental. But it is to pay attention, that reflection is an individual and spontaneously process. Thus reflection is hard to schedule and recognise. Thus self-directed learning is getting significant relevance. The trainer acts as a coach and is more and more responsible for suitable support of the learner. That includes professional counselling, support to organize the learning process and development of social competencies. New media provide a multifunctional help to fulfil these functions.

## 4.3. *Media Resources*

By using new media (e.g. web-based educational materials) and modern communication infrastructures a large amount of information can be carried out almost entirely at the work place and within the real working environment.

Even though instruction remains obligatory in APO-oriented vocational training, conventional instructional resources like books, manuals and static web sites become less importance:

- Contents are getting more unique and very specific; instructional resources to guide and support learning should be conceived that way, too.
- Learning environments are proprietary, many information needed to transform knowledge from theory into practice is company-specific and only destined for internal use. Instructional resources that may accelerate this knowledge transfer must combine domain-specific and company-specific „knowledge".
- Real-work assignments must be solved within reasonable time; e.g. a network administrator must be able to reconfigure a router in a short time. Instructional resources that enable an IT technician to learn about router configuration on-the-job must be made available within hours or even minutes.

These requirements cause a demand for instructional resources that are highly adaptive, highly integrative and can be produced on-demand. Recent approaches which support the on-demand selection and sequencing of modular learning resources (e.g. Caumanns 00) seem to be suited well to provide the flexibility needed for this new kind of instructional material.

## 5. SUSTAINABILITY

Because of the rapid development in the IT-sector, sustainability is a central issue for the quality of vocational training. Therefore, it is not only instruction but also the suggestion that self-learning competencies to up-to-date knowledge effectively on the job is the primary aim of our concept.

It is necessary to find ways to merge informal learning processes and instructive elements.

## 5.1. Fading

An ultimate goal is to foster participants in developing auto-didactic skills and metacognitive competencies. To achieve this, media, methods and the organisational structure of the qualification process should shift from organized instruction to authentic working conditions. That means that the trainer, who teaches basic knowledge at the beginning, gradually supervises and assists the learning process on demand. Participants should organise their comprehension processes, practice their own skills and use tools (e.g. search engines) instead of didactically prepared teaching material (figure 4). This "fading" method is part of a basic approach to develop skills and is oriented to the basic approach of "Cognitive Apprenticeship" (Brown/Collins/Duguid 1998).

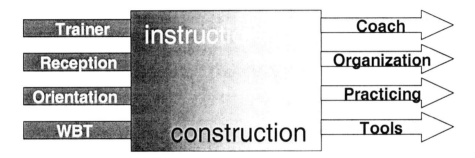

*Figure 4. Fading.*

## 6. PERSPECTIVE

Along with special recommendations for implementation the reference projects are documented in manuals. The manuals enable companies to train IT-specialists independently from producers and provide quality standards.

In addition to on-the-job and vocational training, the modular set-up aims at guaranteeing a smooth transition to the college system. The creation of job profiles and the recommendations generally available as "open source" open up the perspective of using this system even beyond the limits of the educational field.

A first course based on APO will be carried out by the German Telekom AG in summer 2001.

## 7. REFERENCES

APO: http://www.apo-it.de

Albrecht, G. (1999): Personenqualifizierung im Kontext neuer Lernwelten. Bertelsmann-Verlag: Bielefeld

Bitkom (2001): Kein Ende des Mangels an IT- und E-Business-Spezialisten absehbar. Press Release. Internet: http://www.bitkom.net/presse/presseinformationen/060301.html (3.05.2001)

Brown, J.S., A. Collins and P. Duguid (1998): Situated Cognition and the Culture of Learning. In: Educational Researcher, Issue 1, pp 32-43

Bundesanstalt für Arbeit und Sozialordnung– BAM: Informationen zum Fachkräftebedarf in Deutschland. Internet http://www.bma.de/de/aktuell/thema/it1.htm (3.05.2001)

Caumanns, J. (2000): Bottom-Up Generation of Hypermedia Documents. Multimedia Tools and Applications, Vol. 12, Issue 2/3, pp 109-128, Kluwer Academic Publishers, November 2000.

Dehnbostel, P. (2000): Erfahrungslernen in der beruflichen Bildung: Ansatzpunkte für eine neue Lernkultur? In: Dehnbostel, P./Novak, H. [eds.]: Arbeits- und erfahrungsorientierte Lernkonzepte, Bertelsmann: Bielefeld

Grunwald, S. and M. Rohs (2000): Arbeitsprozessorientierung in der IT-Weiterbildung. In: Berufsbildung in Wissenschaft und Praxis, Heft 6, S. 22-24

Kirchhöfer, D. (2000): Informelles Lernen in alltäglichen Lebensführungen, QUEM-report, Issue 66

Livingston, D. W. (1999): Informelles Lernen in der Wissensgesellschaft: Erste kanadische Erhebung über informelles Lernverhalten. In: Quem-report: Kompetenz für Europa: Wandel durch Lernen – Lernen durch Wandel, Heft 60, S. 65-91

Marsick, V. J and M. Volpe (1999): The Nature and Need for Informal Learning. In Marsick, V. J and M. Volpe (eds.):Informal Learning on the Job, Advances in Developing Human Resources, Berret Koehler Communications, San Francisco

Petersen, A. W. and Wehmeyer, C. (2000): Die neuen IT-Berufe auf dem Prüfstand: Erste Ergebnisse der bundesweiten IT-Studie. In Berufsbildung in Wissenschaft und Praxis, 29 (6), 7-12

Staudt, E. and B. Kriegesmann (1999): Weiterbildung: Ein Mythos zerbricht. In: Arbeitsgemeinschaft Qualifikations-Entwicklungs-Management – QUEM [eds.] Kompetenzentwicklung '99, Waxmann Verlag, 17-60

*Matthias Rohs, Walter Mattauch, Jörg Caumanns*
*Fraunhofer Institute for Software and Systems Engineering*
*Berlin, Germany*

J.-C. LÉON, P.-M. BOITEL, Y. DELANNOY

# ENGINEERING WORKSHOPS: A MULTIDISCIPLINARY PROJECT TIGHTLY INTEGRATED IN ENGINEERING EDUCATION

**Abstract.** The engineering workshops designate a global design activity experienced at ENSHMG for more than 10 years. Continuous adaptation and evolution of this teaching activity has led to a multidisciplinary design project supervised by a team of professors having complementary skills and involving communication and economical aspects as well.
After the description of the organizational aspects of the engineering workshops and the review of the main concepts of this activity, its position within the global training program at ENSHMG and its relationships with the topics taught will be addressed. It is aimed at showing how the integration between the engineering workshop projects and standard topics can produce synergies under various forms of the educational environment at ENSHMG.
This activity is part of a progressive project scheme throughout the educational program of the ENSHMG and is now concerning the 150 engineering students at ENSHMG. This paper analyses also the various issues of such an activity in terms of motivation and efficiency for the students as well as the student-trainer relationships and the effects over the in-house educational program.

## 1. INTRODUCTION

The ENSHMG (Ecole Nationale Supérieure d'Hydraulique et de Mécanique de Grenoble) is one among the ten engineering faculties of the INPG (Institut National Polytechnique de Grenoble) and aims at training students in the fields of mechanical engineering and environmental and civil engineering. As such, students are trained over three years.

After an entirely common first year, the students separate according to their department and option choices at the beginning of the second year. The engineering workshops projects take place during this second year. The third year is characterized by student-selected topics and a final year project (4 months) usually spent in industry. Over their training period, all the students spend some time in the industry every year at the end of their academic year.

In the above context, the engineering workshop projects are described and analyzed. This analysis fits within the context of the standard decomposition of training activities, i.e. lectures, practical work, project activities, directed work, in the French educational system and the area of mechanical engineering and environmental and civil engineering, which can be considered as restrictive aspects to the applicability of the current approach to other educative and training contexts.

The following sections cover respectively:
- The training context at ENSHMG in term of position within the French educational system (section 2),
- The detailed description of the organization of the engineering workshop projects (section 3),

- The integration of the engineering workshop projects within the student training scheme (section 4),
- The concluding remarks and perspectives concerning this project activity.

## 2. THE TRAINING CONTEXT

The ENSHMG trains engineering students in the fields of mechanical engineering (according to two main streamlines fluid mechanics and hydraulics and mechanical design and structural mechanics) as well as environmental and civil engineering.

The overall training of the students is strongly based on various in-house project activities taking place during their three years training (baccalauréat + 3 years to baccalauréat + 5 years). During the first year (baccalauréat + 3 years), mainly disciplinary and small projects are undertaken by every student (about 120 students per year), i.e. projects ranging from 30 to 44 hours per student, and a first link is established with the course of project management taught in the same academic year. These projects aim at developing the initiative and autonomy of the students, which is important because they have had no experience in this field up to their entrance into a faculty of engineering.

Two projects are carried out during the second semester:
- One related to strength of materials, which has a duration of 30 hours per student,
- One related to mechanical design, which has a duration of 44 hours per student.

It is not desirable to extend further the duration of such projects because projects incorporating numerous multidisciplinary aspects would be managed with difficulties by the student given their lack of experience in handling project management aspects and technical matters simultaneously.

In addition, the mechanical design project is used as basis to establish the link between project management courses and the project activity itself. This is the first step to prepare the students to larger multidisciplinary activities and it is their first opportunity to organize a project activity with multiple requirements, i.e. project planning and technical objectives. The mechanical design project is also the first configuration where the students are tutored through a team of teachers. In this case, all the teaching staff provides nearly the same type of information to the students because all the teachers have nearly the same background and the level of complexity of the projects does not require specific technical information.

Students are working in groups of two to three students to train them to team work because they have also no experience in this area up to this first academic year at the faculty of engineering. Here, the objective is to get them used to task analysis and task planning so they avoid performing most of the work in common but perform really different tasks.

As it can be seen, this project activity during the first year is a preparation phase to a larger project activity, i.e. the engineering workshop projects, so that the students can gradually incorporate new aspects of such activities while obtaining the largest possible efficiency from the teaching staff point of view.

## 3. THE STRUCTURE OF THE ENGINEERING WORKSHOPS

The engineering workshops are larger projects ranging during the whole academic year and taking place in the second year of training (baccalauréat + 4 years). The duration of this activity is approximately of 150 tutored hours per student. The engineering workshops have been initiated more than ten years ago on a small-scale basis (around 25 students at the beginning of this activity) and progressively tuned and spread throughout all the options and departments of the ENSHMG (reaching now 150 students).

The project themes are close to industrial problems and can be provided by industries, research laboratories or engineering needs internal to ENSHMG. In any case, the students stay at ENSHMG during their project since industrial people are acting as clients or skilled supervisors and get in touch with the students at specific milestones of the project.

Project themes are not necessarily bound to a one academic year scheme and can be spread over several years similarly to industrial projects ranging over several years where engineers are assigned a mission for a part of a larger project. This is clearly a choice to get closer to industrial configurations where engineers can encounter a wide variety of configurations ranging from the early phases to the final stages of a project. In any case, most industrial projects are based on prior work and know-how. Hence, carrying on projects over several academic years generates a configuration incorporating some features of the context of industrial projects.

Thus, this project activity is characterized by a relative complexity of the subjects compared to the projects held during the first academic year. This project aims at being close to a full-scale engineering project. To stay coherent with such a context, the duration of 150 hours appears to be an acceptable compromise between the project objectives and the engineering training goals. Allocating such an amount of time for a project activity can be considered a strategic choice when the overall academic year of a student reaches 850 hours.

Such a context of project activity means that project work among the groups of students differs and ranges between the early project stages to final ones. In order to help the students handle their project, industrial software (CAD/CAM, CFD, structural analysis, ...), measuring equipment and manufacturing equipment is provided together with the help of technical staff. Thus, on-site investigations of physical behaviours as well as manufacture of prototypes can be performed to conform to project needs. Other equipments are also made available from the laboratories used in practical work activities. The cost of such experiments or prototypes is kept as low as possible to stay equivalent to practical work budgets, otherwise negotiations are conducted with the industrial partners to perform on their own budget the corresponding experiments or prototypes and provide only the results to the students.

The engineering workshops projects are conducted by groups of two to three students. The groups are kept small so the students can largely concentrate on scientific and technical aspects rather than organizational ones. This is the pedagogical choice made by the teaching staff of the faculty. In fact, the number of students per group could be extended if the pedagogical objectives were more centred on the project planning and organizational aspects.

In the present configuration, the compromise retained is to organize the students in small groups with loosely interconnected technical objectives to ensure that the behaviour of each group does not act too much upon others while technical objectives are not totally different and may generate some interactions between the corresponding groups. This is clearly a limit set to the interaction of project teams compared to the industrial context but this choice originated from assessment constraints where a large group of students can be hardly evaluated on an individual basis. At the opposite, a small group is usually easier to assess.

To conclude about the structure of the engineering workshop projects, the budget of this activity is similar to directed work activities performed on computers where one tutor is assigned to a group of 16 students and the budget allocated to manufacture prototypes and buy measuring equipments amounts to 4500 € per academic year for the whole projects. At present, the supervision of the projects is performed by a team forming a subset of the teaching staff whose interaction with the projects will be described in the next sections.

## 4. INTEGRATION OF THE ENGINEERING WORKSHOPS INTO A TRAINING SCHEME

### 4.1. Integration with non-technical and non-engineering topics

These projects have revealed the strong interest and motivation of the students who get really involved in this activity. Each project is performed by a group of two to three students to allow them to use project management tools to decompose their whole project into tasks, schedule and spread these tasks between them. Here, the objective is to gain practical experience of project management tools since one milestone together with the final project assessment and continuous supervision are used to enforce the practise of such tools. This is an example of how this activity is used to tie together topics that usually are considered as loosely connected by the students. Project management aspects are based on lectures given to the students during the first year (baccalauréat + 3 years) and progressively put into practise through project activities among which the engineering workshops are the most important one. As stated in section 2, a first level of practise about project management tools is gained during the first year through the engineering design projects. Engineering workshop projects bring more complexity to this practise since the project time is greater and its technical content is more complex. Task time evaluation is difficult for the students and basic uncertainty management is proposed to help the students in their project organization.

Other relationships between disciplines are set up through this project activity in the scientific and technical fields as well as with communication and economical aspects. The integration of communication skills into the project activity is based on a meeting-debate activity to link these disciplines much in the same way engineers do in industry during technical meetings. The students are prepared through group management and meeting management activities with communication specialists. Then, a **co-animated** training between communication specialists and supervisors of the project activity takes place with the students to prepare them to meeting-debate activities related to their project. Co-animation enforces the connection between these disciplines through complementary arguments brought by each co-animator, something that cannot be obtained otherwise given the skills required. Following this training exercise, the meeting-debate activity is inserted into the engineering workshop activity where each group has to identify a specific problem related to their project, to present it to a group of other students and to conduct a debate in order to gain some possible answers to their problem.

From the point of view of economical and financial aspects, integration is obtained through a financial evaluation of each project based on the lectures provided and on the specific aspects of each project. The accompanying lectures are held at the beginning of the second year simultaneously with the beginning of the engineering workshop projects. This activity raises the question of cost of various tasks and equipments used rather than setting a constraint on each project. Here, the goal is to bring a view of the cost of a project rather than to use it as an effective bound. Because the lectures concerning the economical and financial aspects of projects are inserted in the second year of the students' training, the project activity is restricted to simply bridging this topic with the project context and can hardly be developed further since it is only a first experience and also because some elements of costs cannot be evaluated easily.

Such an architecture can only be achieved with long enough project duration. This justifies the scope of the objectives and current duration of these engineering workshops projects.

*4.2. Integration with scientific and technical topics*

From the scientific and technological points of view, the engineering workshops form a specific opportunity for the students to merge together the various topics they have learnt or they are learning through the engineering project they undertake. To this end, the scientific and technical supervision of the projects is achieved using a multidisciplinary team of members of the teaching staff.

Among the particularities of this team, the members own different positions, i.e. professors, assistant professors, senior lecturers, "professeurs agrégés", ... and form a multidisciplinary team. Each of them brings his specific background in one or several fields or topic and acts as an advisor rather than a supervisor and is not dedicated to one or several projects but can bring his knowledge into any project on the request of the students. Currently, the main topics covered by the supervision team in the mechanical engineering field enumerates:

- Structural mechanics,
- Dynamics of solids,
- Design and technology,
- Fluid mechanics,
- Dynamics of fluids,
- Computational fluid dynamics,
- Thermal exchanges,
- Strength of materials,
- Automation and servo control,
- Manufacturing techniques,
- Computer aided design and computer aided manufacturing,
- Numerical simulations and technical data management.

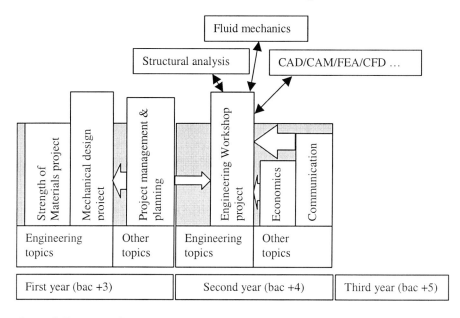

Figure 1. Position and organizational aspects of the engineering workshop projects. The light arrows illustrate the relationships between some engineering disciplines and the engineering workshop projects.

At present, the corresponding team relies on a group of thirteen persons who advise and help over one hundred students involved in this project activity. Such a structure efficiently helps the students, through effective engineering practise, not to consider the various topics as isolated from each other but as real complementary elements of a global solution. The participation of professors, assistant professors, ..., teaching the various topics involved in these projects helps the students not to see these topics as theoretical but real tools or models with their efficiency and limits.

Furthermore, the members of the supervision team appear clearly to the students as being interested in the applications of the topic their teach as well as efficient in or inserting the corresponding knowledge into the context of projects.

Similarly, the groups of students are organized inside each department but the options they have chosen is not taken into account so that this project activity contributes to exemplify how the options can be bridged together and how a project activity can bring together various skills.

As a result, the various disciplines covered by the supervision team are more efficiently bridged with the project activity and gain more interest with respect to the students. Similarly, the relationship between the members of this team and the students is reinforced due to the proximity between both during the project activity.

Conversely, the team members get a more realistic view of the way the students are effectively using the knowledge they bring during their lectures and they can adapt the content of their lectures to improve their efficiency.

A similar approach has been recently set up in the department of environmental and civil engineering, hence the position and organizational aspects related to the engineering workshops projects can be summarized in figure 1. Even though the number of projects tends to be rather large, i.e. nearly fifty different projects for the mechanical engineering department, it has not led to strong difficulties when this activity has been extended progressively from 30 students up to 150.

### 4.3. Assessment of the project activity

The engineering workshop project activity is assessed under the following aspects:
- Project work performed at the first third of the duration (technical aspects and project planning aspects),
- Behaviour during the meeting-debate activity during the second half of the project,
- Financial estimation of the project,
- Final review (oral presentation and report) with a jury.

The first assessment is based on a report and tends to evolve toward a report and an oral presentation to be able to provide rapidly some feedback to the students at that time of the project where groups behaving badly can be reoriented with limited consequences on the project itself.

### 4.4. Benefits of this activity

The multidisciplinary team acts also as a model of the context of real firms where an engineer has to work in a team of engineers and technicians each of them having specific skills. Thus, this project activity acts also as a forum of exchange between the team of professors, assistant professors, … to help new trends to emerge in topics to teach as well as in manners of teaching. As such, this activity can be considered as a place for a melting pot where the culture and habits of a faculty of engineering, considered as an organization, can evolve.

Innovative aspects can be introduced more easily as technology evolves since project themes and organization can be adapted to new requirements. As an example, collaborative engineering tools and techniques has been introduced in the past years through projects themes concerning distant teams of students (some located in Grenoble at ENSHMG and some others located in Annecy (100kms far from Grenoble) in another faculty of engineering). Thus, videoconferencing tools, application-sharing tools, PDM (Part Data Management) systems have been evaluated in a real context of collaborative work. Such activities, together with the knowledge the supervision team can extract from them, help evolve and introduce new topics in the educational training and help spread some know-how across the supervision team since the supervision team gets involved into a significant number of projects.

As a result this activity has been validated within ENSHMG and carry on evolving on a long time scale basis to preserve its efficiency, to help increase the number of participating members in the supervision team. Incorporating new members into the supervision team can be thought as a cultural evolution for professors, assistant professors, ... since they are not used supervise **simultaneously** the activity of students as well as they may feel difficult to define their position when **co-animating** sessions with communication specialists. Indeed, bringing together different grades of teachers (professors, assistant professors, ...) who have restricted or extended know-how in teaching, small or large experience in team work, produce also cultural changes which call for progressive changes rather than rapid ones to keep people confident and really involved in this activity. Thus, divergence of views between members of the supervision team stays natural for the students and do not generate conflicts in this team. All the previous aspects justify the long time scale basis, which is currently necessary to insert and develop such a multidisciplinary project activity within an existing engineering training structure.

As a result, the vision by the students of the professors, assistant professors, ..., members of the supervision team evolves and generates configurations, which are closer to real engineering practise than the formal relationships that are usually encountered. Hence, such configurations can be used to improve the efficiency of topics through examples that can be extracted from current engineering projects.

However, this approach cannot be compared to alternate training of engineering students between industry and education environments since this type of training with long stays in industry cannot offer students the opportunity to perform tasks at an engineering level. At the opposite, the engineering workshops strictly concentrates on merging topics in an engineering context on subjects which can be provided by companies and as such, it takes places in the educational environment.

## 5. CONCLUSION AND PERSPECTIVES

The previous sections have described the engineering workshop project activity as well as its position within the training scheme of the students and its integration within the set of disciplines taught.

The activity set up appears to have a strong impact on the students since it generates significant motivation and help them assimilate the relationships between the various disciplines taught. In addition, the structure set up generates relationships inside a subset of the teaching staff and across engineering as well as non-engineering topics.

The last point is in fact a basis to insert dynamics over time into a training scheme, because the engineering workshop projects can be the place where new technologies can be assessed, where the interest for new disciplines can be rated, where new bridges between disciplines can be set up because members of the supervision team are sharing the same activity. Similarly, new pedagogies can emerge because members of the supervision team contribute to projects in manners that can lead to different ways of teaching some subsets of disciplines. To summarize, the synergy between the members of the supervision team may be also a significant contribution of the engineering workshop projects even more than its effects on the student training.

Among the current perspectives of this activity, it is aimed at taking benefit from this activity to capitalize the know-how about the projects. This task has gain interest because these projects can bring documents, examples, various versions of a mechanism, ..., that can be used in different engineering topics to improve their efficiency and tie them to the reality the students know through this project activity. Such a capitalization task raises interest because intranet facilities and archiving techniques enable to store numerical documents in large quantities and to sort and retrieve documents efficiently using PDM (Part Data Management) software environment. In addition, new technologies can be used when teaching to visualize and get access to numerical data, 3D models, and simulation visualizations.

*Jean-Claude Léon, Pierre-Marie Boitel, Yves Delannoy*
*Ecole Nationale Supérieure d'Hydraulique et de Mécanique de Grenoble*
*(ENSHMG)-INPG*
*Saint Martin d'Hères, France*

R. G. ZYTNER, W. H. STIVER

# CAPSTONE DESIGN

**Abstract.** The School of Engineering, University of Guelph, offers a final semester capstone design course to give graduating students a head start in the real world environment. Students work in teams on projects directly related to the needs of outside clients. Multi-disciplinary projects are encouraged with individual students working on the components of the project that are linked to their discipline. Examples include spine implants, robotic interfaces, stream restoration and membrane-based wastewater treatment. Throughout the project the students liase with the outside contact and regularly meet with a faculty advisor. In addition to meeting technical standards, the project must address social, environmental and aesthetic issues. At the project's completion, the students prepare a written report and poster presentation. Faculty advisors evaluate the written reports and a team of practising engineers from the community evaluate the poster presentation.

## 1. INTRODUCTION

The School of Engineering, University of Guelph was started in 1874, with its roots in Agricultural Engineering. Today the School offers four engineering programs, consisting of Biological Engineering, Engineering Systems and Computing, Environmental Engineering and Water Resources Engineering. These programs are largely interdisciplinary in character relying on different combinations of the fundamentals from the more traditional engineering fields of Civil, Chemical, Electrical and Mechanical. Thus, the School's offerings are quite unique in Canada. The School's Engineering programs are also differentiated from the others through a heavy emphasis on team and project work, a greater integration of life sciences and an undergraduate population that is 42% female.

All four programs are accredited by the Canadian Engineering Accreditation Board. This is important for Canadian students as it allows them to become a licensed Professional Engineer anywhere in Canada, following four years of practise under a licensed Professional Engineer, provided they pass a Professional Practice Exam (PEO, 2001; the number of years of required practise does vary from province to province and the nature of the Professional Practice Exam also varies). No further technical examinations are required to practise engineering in Canada. It should also be noted that under the North American Free Trade Act, which includes Canada, Mexico and United States of America, there are ongoing discussions to have reciprocating agreements with these jurisdictions.

An interesting component of the School is that it is not departmentalised. This provides many positive features, including a common design sequence for all engineering students. The design sequence consists of four engineering courses, one in each year: Engineering & Design I, Engineering & Design II, Engineering & Design III and Engineering & Design IV. Each course in the sequence adds to the complexity of the design material taught the previous year. Engineering & Design IV (our Capstone Design course) is the culmination of all the engineering and design skills that the students have acquired. It is taken by the students in their final semester and is delivered in a project-based learning format. This paper provides detail on the design sequence and outlines how the Capstone Design course is delivered.

A new development for the design sequence is the Canadian Design Engineering Network (C-DEN). C-DEN links all Canadian Engineering Schools electronically and is used to share design information across the country. C-DEN's planned involvement in the design sequence will also be highlighted.

## 2. DESIGN SEQUENCE

The School of Engineering takes pride in the Design Sequence that is offered. It offers both multi-disciplinary and discipline specific projects, all developed so that the students acquire the maturing design skills. All the courses in the sequence are team based. The following sections outline the contents of each course.

*2.1. Engineering & Design I*

Engineering & Design I is taken by all engineering students in their first semester. The approach is multi-disciplinary, such that students from each program are intermixed in the teams. There is introduction to engineering and design by means of selected problems. The students integrate basic science, mathematics, and complementary studies to develop and communicate engineering solutions for simple engineering design problems. The design problems tend to have a little dimension of fun to them. An example from previous years was the design and building of small sailboats using a finite list of recyclable materials. Evaluation of the sailboat designs was based on their speed, which was specified at the beginning as the key design criterion. Applications of computer-aided drafting, spreadsheets, programming, and mathematical solving packages are also incorporated into the course.

*2.2. Engineering & Design II*

Engineering and Design II is the second course in the sequence and is taken by all students in their third semester. The projects are again multi-disciplinary. Last year, one of the projects was to design a reuse for all components of one million personal computers. The project started with time in the machine shop pulling apart a complete computer, identifying and weighing all of the parts. The course includes a particular emphasis on individual oral communication skills. Use of computers in engineering and design is further developed with the integration of the use of database software, high-level computer programming, and computer-aided-engineering software.

## 2.3. Engineering & Design III

Engineering & Design III is taken in the sixth semester of the student's program. This provides sufficient exposure to engineering science courses in the student's program, allowing them to tackle discipline specific open-ended problems. Examples include portable water treatment system, audio signals for a pedestrian cross for visually handicapped people, electronic transmission of medical measurements from home to the doctor's office, removal of methane from a landfill and the biological treatment of hospital waste. Additional design tools are also presented in the course, including model simulation, sensitivity analysis, linear programming, knowledge-based systems and computer programming. Complementing these tools are discussions on writing and public speaking techniques, codes, safety issues, environmental assessment and professional management. These topics are taught with the consideration of available resources and cost.

## 2.4. Engineering & Design IV

Engineering & Design IV consists of four courses; a semester long course for each program offered by the School of Engineering. Students register for the course identified in their program of study. A team project is selected, which involves the application of engineering principles and computers to design a solution. Each team prepares reports and presentations to professional standards. Even though each course is discipline specific, students are encouraged to inter-mix, providing their respective strengthens to the project they are involved in. Further detail on course delivery is provided in the remaining sections of the paper.

## 3. DELIVERY OF ENGINEERING & DESIGN IV

All members of the faculty deliver Engineering & Design IV to the students. One faculty member is assigned the co-ordination responsibility, which covers team formation, advisor selection' deadlines, grading weighting and formats, arranging the poster presentation day as well as selecting the poster evaluation panels. The specifics of each task are outlined below.

## 3.1. Formation of Team

Students are encouraged to work in teams of 3 or 4; however, on occasion teams of 2 and 5 are permitted. They are also encouraged to consider cross-disciplinary teams. Team formation is student driven. The keen students form teams as early as a year before the course is taken, whereas some leave it to course registration time (approximately 2 months ahead). All students registering for the course are required to make prior arrangements with a team. The students generally select their teammates based on a combination of friends, individuals that they are comfortable working with from past projects and common interest in a domain or topic for their design project.

*3.2. Project Theme*

The students generally take the lead in finding a project. Most teams are able to use their existing contacts (through summer or co-op employment experiences) or establish new connections to find an outside problem that fits both their interests and the requirements of the course. Faculty advisors typically play a role at this stage in helping the students and the outside client establish a realistic scope in the context of the course. Increasingly, outside clients come to us offering problems that may be suitable for student-based design. The project theme must be established at the time the students register for the course. The projects worked on by the various teams are either preliminary or final designs. Where appropriate, prototypes are constructed. Table 1 gives examples of the completed design projects for the Winter 2001 semester.

*Table 1. Capstone Project Titles in Winter 2001*

| Program | Project Title |
|---|---|
| Biological Engineering | 1. Mono-Loc: A Suture Tightening Instrument for Canine Knee Surgery<br>2. Edible Soy Packaging |
| Engineering Systems and Computing | 1. Intelligent Mechatronic Aide for Recovering Stroke Patients<br>2. Design of Workstations for Teaching Robotics |
| Environmental Engineering | 1. Preliminary Design of a Wastewater Treatment Process for the Village of San Francisco, El Salvador<br>2. Design of VOC Emission Reduction Strategies for Polycon Industries in Guelph |
| Water Resources Engineering | 1. Low-flow Hydroelectric Production on the Grand River<br>2. Integration of Stormwater Management Facilities with GIS in Oakville |

*3.3. Role of Faculty Advisor*

Typically, faculty members advise up to three projects per year in their area of expertise and meet with each of their design teams at least once per week. In addition, all faculty and staff of the School are available to all of the design teams for advice. Faculty advisors play an important role in the how the project evolves with their role responding to the needs of the team. At times, faculty act as the principal of a design firm, at times as the client, at times engaging in discussion of some detailed calculations and at times acting just as a sounding board. However, it must be stressed that in all cases the students have ownership and responsibility for their designs and their design decisions.

### 3.4. Grading of Projects

The team's design project grading consists of a proposal (10%), interim report (20%), final report (50%) and poster presentation (20%). Faculty advisors are responsible for grading their team's proposal, interim, and final reports using a format that is common for all design teams.

The poster presentation has recently been implemented as the presentation form and it has become an excellent vehicle to promote the projects and the School of Engineering. The poster presentation takes place after the last day of classes, just as if it was an exam. The presentations take place in the University Centre, where there is high visibility for the entire university community. Many positive comments have been received as to the type and quality of projects. Grading is open to all visitors, faculty advisors and practising engineers invited by the course co-ordinator.

### 3.5. Challenges

The School of Engineering is proud of the way the Design Sequence and the Capstone Design develops the design capabilities of our students. However, there are challenges and areas for improvement. In Capstone Design, there are few areas that need to be discussed.

A challenge with a team-based approach is to ensure that every individual in every team participate actively in the team's design and strengthen their individual design skills. For the vast majority of the students and teams this is not a problem. Most often they are working with friends and colleagues and the sense of commitment to each other is sufficient to keep all students active. However, every year there are a small number of students whose activity raises concerns. To address this issue, students are asked to evaluate their contribution and the others in the course. In the majority of cases, the students report that all members contributed equally. However, having this mechanism in place does provide a means for adjusting the grade if a situation develops.

Inconsistency in depth and challenge in the various projects is a potential concern. The freedom to select problems from outside clients and the diversity of problems that are chosen make complete consistency impossible. Faculty advisors are charged with helping to establish the project's scope and to avoid extremes at either end of the challenge spectrum. However, with up to 25 faculty members involved complete consistency is not a realistic expectation. Alternative approaches that have been discussed to improve consistency include involving fewer faculty members in the advisory role or creating the same or similar design problems for each team. The former solution has been rejected on the basis of workload distributions and ensuring that a breadth of expertise is available to the teams. The later solution has been rejected on the basis that the student's selection of their projects to suit their interest leads to greater enthusiasm and engagement in the project than would otherwise be possible.

The grading of the various design team's work is inconsistent. Even though a common grading guideline is followed, every faculty member interprets these guidelines differently. The grades differ as a result and can differ by significant amounts. The students are the group complaining the loudest on this issue. Some see their colleagues having an easier time and yet getting a higher grade. Some faculty advisors have reputations as easy and some as being tough. Some faculty argue that this inconsistency is no different from the working world model, where the pay scale varies from company to company. Alternative solutions include evaluating the projects on a pass/fail basis only or again to reduce the number of faculty advisors involved. The former may foster students to focus on what is the minimum they need to do to pass and lead to lower quality products overall. The later solution again creates the workload difficulties and the availability of expertise.

The last significant challenge involves professional liability. As indicated earlier, many of the projects are real world problems. In many of these cases, the person providing the project theme is looking for a report outlining the results of the design so that it can be implemented. The problem is that a professional engineer did not develop the engineering project. While the majority of faculty members are professional engineers, faculty are not reviewing the project from a consulting engineering perspective. Instead it is an academic exercise. To inform the clients, who are not charged for the service provided by the project teams, that the project was not design by a professional engineer, a disclaimer clause is added. This clause indicates that the work belongs to the students and that the design must be reviewed by a licensed professional engineer before being implemented. To date none of the project clients have had a concern with the disclaimer.

## 4. CANADIAN DESIGN ENGINEERING NETWORK

A recent change in the Canadian engineering education is the development of Canadian Design Engineering Network (C-DEN). This network is being implemented by 31 of 32 engineering schools in Canada using the world wide web. Funding for the network's start up is from the engineering schools and NSERC of Canada with the goal of long-term funding from industrial design projects.

The goals of C-DEN are to strengthen engineering design education in Canadian universities and to foster increased research in the design field. The sharing of instructional design ideas and design expertise through the network is the primary means to achieve these goals. Course modules and projects are being developed as web-based resources to share between institutions.

Each university has a local C-DEN committee who ties into the larger network. In Guelph, the committee comprises the design sequence instructors, which provides an ideal format for the C-DEN initiative to positively influence the courses in the design sequence and specifically the Capstone Design course.

A final component of the network is the establishment of 15 NSERC Engineering Design Chairs distributed across the country to help foster both local activities as well as networking activities. The School of Engineering has been awarded an NSERC Environmental Design Engineering Chair.

## 5. SUMMARY

The School of Engineering has developed an innovative way of incorporating design into the engineering curriculum, the Design Sequence. It consists of a design course in each year that is taken simultaneously by all four engineering programs. The courses offer both multi-disciplinary projects and disciplinary specific projects. By completing the sequence, students develop their design skills.

An integral component of the Design Sequence is the Engineering & Design IV course, otherwise referred to as Capstone Design. This course is taken in the final semester of study and requires the student to work on a project in their program of study. It culminates with a poster presentation to peers and practising engineers.

A new development in Canadian engineering education is C-DEN. This network links all but one engineering school via the web, through which students and faculty can obtain important design information. This will strengthen the design skills students will have before they leave school.

## 6. ACKNOWLEDGEMENTS

The authors would like to thank all the students and faculty at the School of Engineering, with out whom a successful design sequence could not be run. More information on the School of Engineering can be obtained from www.eos.uoguelph.ca.

## 7. REFERENCES

PEO (2001) Professional Practice Exam, Professional Engineers Ontario, Toronto, ON Canada, www.peo.on.ca

*Richard G. Zytner, Warren H. Stiver*
*University of Guelph*
*School of Engineering*
*Guelph, ON, Canada*

W. W. DURGIN, D. N. ZWIEP

# GLOBAL PROJECTS PREPARE WPI STUDENTS FOR THE 21$^{ST}$ CENTURY

**Abstract.** The WPI project based curriculum, which emphasizes discovery based learning are an alternative to the traditional information transfer process, has proved successful in delivering global engineering education. More than 25% of the learning process of the students is integrated into two formal projects, the Major Qualifying Project (MQP) which is designed as a capstone for professional technical competence and the Interactive Qualifying Project (IQP) which relates science and technology to societal concerns and student needs. Both the MQP and IQP may be completed on- or off-campus. Currently, over 50% of the graduating class will have completed one of the projects at an overseas location under WPI's Global Perspectives Program. Each year, more than twenty faculty members will be advising and sharing a learning experience with the students at international locations spanning six continents.

Living and working in an unfamiliar culture while pursuing real world problems of importance to local agencies or organizations provides a unique and stimulating learning environment. Students are fully immersed in the local culture and conduct their studies under the guidance of WPI faculty members. Traditionally, global projects have emphasized the inter-relationship of technology and society through the IQP. More recently, technical projects and research have been added through the MQP and graduate research efforts. The result of the student projects which are generally carried out by small teams, 3-5 students per team is typical, includes oral presentations and a final written report which is presented to a sponsoring agency as well as filed for future use at the WPI library.

This paper describes the WPI global program and is based on the experiences of the authors in advising project activities. It emphasizes the preparation of the WPI students for global projects, the infrastructure needed to support such activities, and the outcomes in terms of global aspects of some graduates' careers.

## 1. INTRODUCTION

Helping students to learn to learn for themselves so that they are better prepared for a professionally competitive environment has been the hallmark of the WPI programs since 1970. In teaching the students to learn for themselves project activities that rely on self-directed study have proved effective. These are called the Major Qualifying Project (MQP), the Interactive Qualifying Project (IQP), and Humanities Sufficiency. Furthermore, because the daily commerce of countries now takes place in a global context, the graduates of the technological universities, such as WPI, must be fully prepared to accept a new type of leadership role, one that integrates technological and humanistic-social know perspectives that can be applied in a variety of global situations.

WPI has been a pioneer in international projects by enabling students to travel to residential sites around the world where they work as junior professionals working on problems defined and sponsored by international corporations and local sponsors. This process provides a distinctive, perhaps unique, professional learning experience for the students. Typically, each student earns credit for the equivalent credit of three courses (nine semester credit hours). In addition. most students going overseas enroll in a pre- qualifying project (PQP) in which they study the country and its culture, gain a rudimentary knowledge of the language of the country, and gather preliminary data for their project. The PQP is the equivalent of one course. The essence of the global perspective program is that of living and working on real-world problems while immersed in the local (unfamiliar) culture.

WPI's first global project center was opened in Washington. D. C. in 1974; its first overseas in London in 1987, and presently conducts programs at 20 overseas locations.

## 2. OBJECTIVE

This paper provides a synopsis of the global project program at WPI and the utilization of WPI's distinctive degree requirements, the IQP and MQP, in preparing engineering students for the world-wide practice of engineering in the 21st century. Project experiences of the authors in England, the Netherlands, and the United States will be used as examples. MQP, IQP; The interactive Qualifying Project, IQP, is usually completed in the junior year and requires students, usually teams of three to five members, to identify, investigate, and communicate effectively on a problem at the interface of science or technology and society. The Major Qualifying Project, MQP, is generally completed in the senior year and requires students to integrate and utilize their prior academic work to solve problems or perform tasks in their major field of study with in-depth understanding and confidence. Effective communication of the results through written reports and oral presentations is mandatory for both the MQP and IQP. Both projects require a minimum involvement of one-fourth of an academic year of full time study. For engineering students The MQP is often used to meet the "capstone design" requirement of their program. To date, about half of the IQP's are being completed by students in residence for eight weeks at overseas sites. The implementation of MQP activity at an overseas location has become more difficult but is increasing and is usually done in a semester abroad, 16 weeks, or partially at WPI.

## 3. DEVELOPMENT OF GLOBAL PROJECTS

In a rapidly increasing global context of business, industry , education and government, the graduates of WPI are being prepared to work and relate with people and cultures on a world-wide basis. At WPI the Global Perspectives Program which began more than a quarter of a century ago is designed to enable its undergraduate students and faculty to become personally involved with the cultures and customs of unfamiliar workplace environments. At present, 1999-2000, more than half of the junior class registered for projects abroad and were accompanied by about 20 different faculty members. To date, the IQP has been the principal mechanism for study abroad: a limited but growing number of students undertake MQP's both abroad and at residential domestic sites.

The IQP and MQP are required academic degree requirement for each student at WPI. The range of possible topics for projects is virtually unlimited as a project is a problem in need of a solution, of special interest and priority to a student or a student team, requires professional competence to solve and there is seldom, if ever, a unique solution.

Both the IQP and MQP may originate with students, faculty, or with external agencies and must focus on solving actual problems associated with business, industry, government, social and educational organizations. The fundamental elements of any project whether on-campus or at a foreign location include a problem description, designation of the student team members and faculty advisor, and statements of the resources available and required. Financial sponsorship of projects whether locally or at one of the global project sites is strongly encouraged but must follow WPI's policies that ensure separation of work for hire from academic work. Sponsoring organizations must understand that the final report and recommendations are those of the students and not of the faculty as consultants.

Students usually get started on a project, MQP or IQP, by reviewing the Project Opportunities web page, meeting with faculty advisors, discussing recent projects or proposed projects with the Project Office Staff, and establishing a continuous dialog with students who are currently involved or have precursor projects. A college-wide Projects Planning Day takes place in April and a Global Projects Fair designed for students desiring to complete a project at an overseas location, is held in October.

A faculty advisor for an IQP may be from any disciplinary area, technical or non- technical at WPI. For the MQP the faculty advisor must be from the disciplinary area of the student. For students who carry out an off-campus project, domestic or overseas, a Preliminary Qualifying Project, PQP, is required prior to the off-campus term. Students report that they spend 45-60 hours per week on an off-campus project. Written and oral reports of a professional quality are required. When appropriate, video or audio tapes may be a portion of the report. The Writing Resources Center and the Instructional Media Center are available for the improvement of a student's oral and written communication skills. WPI places a high priority on excellence with regard to presentation, proposals, reports, and abstracts associated with projects.

The academic advisor assigns the grade of each student on a project and reports the grade to the Registrar through the use of the "Completion of Degree Requirement" form. The final report is distributed within a sponsoring organization by the advisor or project liaison individual. Project abstracts are published annually on the web site. Projects are catalogued for reference in the WPI Library, are retained for five years, and are available on request to interested individuals.

Legal agreements, insurance, and similar documents/activities for students and organizations are normally handled through the WPI Projects Office or the Interdisciplinary and Global Affairs Office for international projects. Typical examples are student housing, financial arrangement with sponsors, and the use proprietary data.

## 4. PRELIMINARY QUALIFYING PROJECT, PQP

The PQP my be directed by the WPI project advisor or by faculty with specialized expertise in the project or the project location. Included are such items as project proposal preparation, teamwork, subject area, background research with preliminary data collection, and individual and group meetings with persons who can brief the students on expected cultural changes, language, location, and monetary differences that are relevant to their projects. For students registering for an IQP at WPI emphasis is often placed on utilizing their prior studies in the humanities and social sciences. For students planning an overseas project the PQP is normally the equivalent of one course: for a domestic project one half of a course.

## 5. TYPICAL PROJECTS WITH GLOBAL IMPLICATIONS

### 5.1. London, England

Challenge: The RAP Museum in Hendon needed to get repeat visitors by gaining the interest of younger, non-WWII population.

Solution: The student teams developed an oral presentation coupled with an involvement of the visitors in a theory of flight demonstration which could be used by the museum volunteer guides. The Museum followed up the success of this project by setting up a major portion of the Museum for "hands on" involvement by the visitors.

Sponsor: The RAP Museum at Hendon, England

### 5.2. London, England

Challenge: The workers, all visually impaired, at the London Association for Blind People needed to improve the efficiency and safety in manufacturing notebooks and similar non-metallic items used in banks. Low worker morale was a concurrent challenge.

Solution: The students prioritized a number of answers and then joined the workers, after agreeing to work blindfolded, to jointly develop answers to both the efficiency and morale problems.

### 5.3. Delft, the Netherlands

Challenge: Delft University of Technology students designed a support structure for a "Floating City for 10,000 People." The WPI students were to develop an operational "topside" to the structure.

Solution: The WPI team developed a model for the viable operation, physically and financially, of the City including the governance system, rental costs, utilities, and emergency services, for example. The study confirmed the excellence of the results of students of different backgrounds and languages who, through an integrated global program, established a model as students for later work as professionals.

## 5.4. Stow, Massachusetts, USA

Challenge: The Collings Foundation in their restoration of a WWII Grumman Avenger needed to develop a history of the particular airplane along with some physical help in the restoration project.

Solution: Student teams working at WPI and at Mayocraft where the Avenger was being restored evaluated and examined existing literature, conducted interviews with recognized experts in the history of the aircraft and aircraft restoration, worked with volunteers on the actual restoration of the airplane, and made recommendations for the identification and development of WWII technologies associated with the Avenger, and the results of these technologies on the outcome of WWII.

Sponsor: Mayocraft and the Collings Foundation

## 6. STUDENT OBSERVATIONS

Following the completion of a professional level project WPI students universally report that they now feel confident they can successfully compete in the international market place associated with their professional, technical fields. They have developed an ability to work in another culture as well as communicate and interact with persons whose native language is not English. The non-traditional element of the learning experience has developed in each student a degree of confidence and expertise well beyond the traditional classroom and textbook learning processes. Students state they are confident they can adjust, work and live successfully virtually anywhere in the world. It is especially important that living and working at a professional level in an unfamiliar cultural setting provides unsurpassed opportunity to gain a solid foundation in "global engineering."

Student evaluations of the faculty advisors and the programs are almost universally outstanding in their praise. Recruiters for the employment of college graduates are among the strongest supporters of the Global Perspectives program and the students who have been associated with it.

## 7. QUALITY ASSURANCE

Three basic words, people, programs, and facilities, are the key elements of success of domestic and global academic programs.

1. Students, faculty, and staff prepare themselves for a global project experience through the development of interpersonal skills, coupled with understanding and recognition of the people and the culture where they will be located. They establish communication with advisors, mentors, and other who may be related to their projects.
2. The fundamental program of preparation for overseas involvement in project work is the non-traditional WPI Plan. In addition there are many supporting programs such as campus seminars, language courses, discussions with natives of the country where the project will take place and contacts with international professional societies, for example.

3. Facilities associated with global projects are highly varied. Where possible, students live under circumstances similar to their counterparts in their new country. Arrangements for housing and living arrangements as well as an academic support area are completed prior to leaving WPI. All of the students must successfully present oral and written documentation of their project prior to returning to WPI.

## 8. FACULTY ADVISORS

The faculty advisor is the key individual at WPI's global project sites. A systematic evaluation of the faculty views of the personal and professional trade-off involved in advising at overseas locations is being conducted by the WPI Provost's office. The evaluation process involves several phases including discussions with faculty participants and identification of issues, determination of focus groups to conduct campus-wide surveys, and the analysis of results and policy implications. Some of the issues already under review include the rewards for faculty commitment as teachers to project advising, the rewards to faculty as professionals judged for merit raises, promotion, and tenure, the impacts on the research programs of the advisor, and the type of short- and long-term preparation of the faculty for global project work.

## 9. CONCLUSIONS

The students and faculty who have been involved with global projects universally state that the educational and cultural experiences have been personally and professionally rewarding to them and should be continued and expanded. In this regard, the Global Perspectives Program has been given increased emphasis and made a major thrust of WPI's current capital campaign. The excellence of the global projects has reinforced WPI commitment to projects as a fundamental part of the teaching-learning process. Student have successfully utilized their required MQP and IQP as a mechanism for learning to learn for themselves in a professionally competitive atmosphere and in that way have distinctively prepared themselves for professional careers in the 21st century.

## 10. REFERENCES

This is not a bibliography. Rather, it notes sources to which specific reference could be made as well as sources where help was less tangible.

1. The Christian Science Monitor, "Today the Classroom, Tomorrow the World", March 7, 1994
2. ASEE Prism, "Globalization of Engineering Education", April, 1995
3. Memorandum of Agreement, Worcester Polytechnic Institute and technical University at Delft, Faculty of Civil Engineering, 19 act 93.
4. Zwiep and Massie, "Educating Engineering Students for a Global Social Responsibility", 1995 World Conference on Engineering Education, act. 15-20, Saint Paul, Minnesota
5. Durgin, Jamieson, Schachterle, and Vaz, "Impacts of Advising Projects Abroad", Second Global Engineering Workshop, Crystal City, Virginia, November, 1988

6. Zwiep, D. N., "WPI: A Leader in the Implementation of Academic Change", A presentation at DUT, 15 June 94
7. Catalogs of DUT and WPI, Academic years of 1998-99,1999-2000.

*William W. Durgin, Donald N. Zwiep*
*Worcester Polytechnic Institute*
*Worcester, MA, USA*